Natural Gas:
Governments and Oil Companies
in the Third World

Natural Gas: Governments and Oil Companies in the Third World

ANN DAVISON
CHRIS HURST
ROBERT MABRO

Published by Oxford University Press
for the Oxford Institute for Energy Studies
1988

Oxford University Press, Walton Street, Oxford OX2 6DP
Oxford New York Toronto
Delhi Bombay Calcutta Madras Karachi
Petaling Jaya Singapore Hong Kong Tokyo
Nairobi Dar es Salaam Cape Town
Melbourne Auckland
and associated companies in
Beirut Berlin Ibadan Nicosia

Oxford is a trade mark of Oxford University Press

OIES books are distributed in the United
States and Canada by PennWell Books, Tulsa, Oklahoma

© Oxford Institute for Energy Studies
1988

British Library Cataloguing in Publication Data
Davison, Ann
 Natural gas : governments and oil companies
 in the Third World.
 1. Developing countries. Natural gas industries
 I. Title   II. Hurst, Chris   III. Mabro, Robert
 338.2' 7285' 091724

ISBN 0-19-730008-1

Typeset by Oxford Computer Typesetting
Printed in Great Britain
at the University Printing House, Oxford
by David Stanford
Printer to the University

PREFACE

When gas is discovered in a developing country, and there is either insufficient to justify an LNG export project, or a surplus over and above LNG requirements, what are the problems that hinder its development for the internal market in that country? Are there positive steps that can be taken to facilitate such development?

Any development of a gas reserve of course presupposes the existence of suitable markets for the gas, and the availability of capital to finance the field development and infrastructure. We therefore analyse in this study the extent and the limits of opportunities for gas utilization. But within the boundaries set by these parameters, a successful gas development scheme depends on two main protagonists reaching agreement over ways and means.

These protagonists are likely to be the government of the country concerned and an international oil company. Natural gas is ultimately the property of the state, which has ownership rights over its exploitation, and hence expects to earn some of the economic rent released by its utilization. A government-owned body is also the usual customer for gas in developing countries. Oil companies normally explore for and discover gas, and have the appropriate technical skills to exploit it, in return for which they expect to make a profit. In some countries the national oil company may operate in its own right: but as a separate institution from the government it will often have very different aims and objectives to the government as such. The major focus of our study is the problems that arise in negotiating and implementing agreements between companies and governments. The asymmetries and differences between the behaviour and perceptions of the two groups impinge on the conduct of negotiations and the nature of agreements reached between the parties. For each group we therefore examine their objectives, the procedures they follow and the constraints under which they operate. We look in detail at the effect of such differences on exploration contracts, on pricing and on fiscal regimes; and attempt to suggest practical ways in which the different objectives of governments and companies can be reconciled to their mutual advantage.

The book is divided into two parts. Part I contains a synthesis of our views on the issues raised by our research, and our main conclusions. It is thus the heart of the study. Part II consists of a series of case-studies of eight developing countries, all of which have natural gas but have developed it to varying levels. The countries included are: Argentina, Egypt, Malaysia, Nigeria, Pakistan, Tanzania, Thailand and Tunisia. These studies are based on a combination of visits to the countries concerned (in the case of all except Pakistan), interviews with oil company executives in Europe and the USA, and desk research. They provide the detailed evidence from which our generalized conclusions are derived.

Many other developing countries of course also have gas reserves, but we had to choose a sample for the purposes of our study. This group of countries was chosen in consultation with the sponsors of the study, and according to two main criteria. First, we wished to include countries which had reached different stages in the development of gas use, in order to illuminate the influence of particular economic junctures on their success or failure. Thus Argentina and Pakistan both have long histories of gas use, and have well-developed markets; at the other extreme Tunisia and Tanzania are both still in the very early stages of implementing gas programmes. Secondly, we felt it was important to include a geographical spread in our sample, with at least one country from each of the major world regions.

Running through all the case-histories is the importance of defining a clear market for the gas. In some cases this can prove remarkably difficult, especially when the oil price is relatively low. In other cases a market does exist, but is very limited in relation to the size of available reserves. The other theme which recurs over and over again is the importance of the relationship between the government and its agencies, and the foreign oil companies which are involved in exploration and development of gas reserves. These two issues are addressed in detail in each case-study. But it is also the case that each country highlights specific aspects of the gas story. The most outstanding are as follows.

Argentina has large gas reserves, and a well-developed and diversified market for the gas. Its problem lies in transporting the gas to its market. Pipelines emerge as difficult political animals, as well as highly expensive tools. For they bring great economic opportunities to those regions they pass through, while closing off such opportunities for the areas they bypass. Raising finance to build them is also a major problem for most developing countries, highlighted in the case of Argentina, where problems of finding a market are negligible.

Egypt is a good example of an oil-producing country where interest in gas came almost as an afterthought. There is no insuperable market problem; but a coherent government policy for gas has been lacking. Even when policies have been formulated they have not been flexible enough to cope with changing market circumstances. Also, as in other countries, oil companies are faced with a supply-constrained economy, with a large and worsening debt problem, which leads to great worries over the convertibility of earnings from an internal gas market.

Malaysia has been successful in promoting gas export projects, in the form of LNG, and is now hoping to develop gas for the internal market on the back of a pipeline exporting gas to Singapore. The project has however been fraught with difficulties, due to internal differences between the various states, and between different government agencies.

Nigeria missed an earlier opportunity for an LNG scheme, and after many years of trying to solve the problem still finds itself flaring enormous quantities of associated gas. While the internal market could never absorb all the gas which is currently being flared, there is undoubtedly further scope for market development. Severe problems in government planning processes and bureaucratic inefficiencies have stood in the way of greater progress.

Pakistan is a case of arrested development. As early as the 1950s a combination of favourable economic circumstances, and a very good relationship between the government and Burmah Oil, allowed for the development of gas. But progress was halted at a later date, due to low prices, and a lack of incentives for further exploration.

Tanzania has small, but significant reserves in relation to the country's size. Its problem is the restricted internal market, and lack of capital to finance infrastructure. For such a poor country however the prospect of using the gas is very tantalizing, and efforts have continued to find an export project which could provide a reasonable return on the resource.

Thailand demonstrates the problems that can result from delays in negotiations, and lack of clear government direction. The government has been keen to encourage gas utilization, but its internal prevarications have worked against the success of such plans. At the same time gas production in Thailand is relatively expensive, and there is strong competition from alternative fuels – local lignite, fuel oil from Singapore, or coal from Australia. This makes its gas market extremely sensitive to movements in international fuel prices.

The *Tunisian* government has been willing to provide incentives to oil companies; but it has met with problems because its reserves are mostly small. It therefore needs to carefully match reserves to small well-defined markets, calling for case-by-case solutions. Meanwhile the fall in the oil price has had special effects on Tunisia, which has access to cheap spot cargoes of fuel oil to meet its limited demand, as well as Algerian gas which passes through the country.

We hope that this study will be of practical significance: that it will allow the development of gas reserves which would otherwise have lain idle; and also that it will help to clarify the conditions under which it is appropriate to try and develop such resources.

ACKNOWLEDGEMENTS

This study was carried out thanks to the financial support of a number of sponsors, namely: Arab Banking Corporation, British Gas/ UK Department of Energy, British Petroleum, Dubai Aluminium, EEC Commission, Elf-Aquitaine, ENI/AGIP, Exxon, Foster Wheeler, Gaz de France, OPEC Fund, Petro-Canada, Ruhrgas, Shell International, Statoil, US Department of Energy, and the World Bank. In addition to financial aid these sponsors were most helpful, providing detailed information, and offering numerous opportunities for us to discuss ideas with members of staff. The sponsors however do not necessarily agree with, or endorse, all the views expressed in this study. In fact each may disagree with particular views, conclusions or policy recommendations.

The study depended heavily on interviewing those directly involved in the processes of negotiation and policy formation, drawn both from sponsors and other sources. We held extensive discussions with government officials in a number of developing countries, with oil company executives, with bankers and with staff of multilateral aid agencies, especially the World Bank. These are too numerous to name individually, but we would like to record our thanks to them for sharing their knowledge with us. Our desk research and interviews were also complemented by a series of steering meetings in Oxford. For these meetings the sponsors of the study were joined by high-level representatives from a number of developing countries, namely, Mr Mustapha Haddad from the Entreprise Tunisienne d'Activités Pétrolières, Mr Rastam Hadi, Managing Director of Petronas in Malaysia, Mr Tongchat Hongladaromp, Governor of the Petroleum Authority of Thailand, Mr Al Noor Kassum, Minister of Energy and Minerals for Tanzania, and Dr Moustafa Hamed Moustafa, Vice Chairman for Natural Gas with the Egyptian General Petroleum Company. These meetings and interviews were invaluable in forming our views, and we are most grateful to all those who freely gave of their time and experience. Needless to say, all views expressed in this book are ours alone, and should not be attributed to any of the very wide variety of people who have shared their opinions with us.

The core research team on the project consisted of Mr Robert Mabro, Ms Ann Davison and Dr Chris Hurst. Ms Rosemary Thorp, of St Antony's College, Oxford, carried out a case-study of gas development in Argentina, and wrote a thought-provoking paper on the relationship between international companies and governments in developing countries. In addition Ms Penny Walker carried out many interviews with oil companies, banks and financial organizations in the USA. Mr Peter Beck provided advice on the conduct of the project and criticism of successive drafts of the report. Dr Margaret Chadwick wrote a market study on methanol. Mr Graham Harris, of BP, spent six weeks with us in Oxford, during which he wrote a market study on ammonia/urea, and also a piece on non-recourse financing from an oil company's perspective. Dr Michael Klein, from the World Bank, contributed a number of papers towards the study, and also discussed the issues at length with us. Dr Keith Palmer, of Rothschilds, wrote a paper for us on the financing of gas projects as seen from the banker's point of view. Furthermore a number of experts gave valuable comments on the final report, most notably Mr Jim Jensen, Dr Aman Khan, Professor Edith Penrose and several members of the Energy Department at the World Bank.

The task of writing the final report of the study, a synthesis of the research carried out by the core team and other contributors, was undertaken by Ann Davison and Robert Mabro. Robert Mabro was mainly responsible for the text of the Main Report, now Part I of this book, and Ann Davison took full responsibility for writing up the country studies incorporated in Part II. She also edited both parts ensuring that the book would read as if entirely written by a single pen.

Working papers which provide support for the analysis or examine in greater depth particular issues are available on request from the Oxford Institute for Energy Studies (29 New Inn Hall Street, Oxford).

Finally, thanks are due to Margaret Ko and Susan Millar for their patient work on the word processor, to David Guthrie for organizing the production, to Karen Exley for arranging the sponsors' meetings, to Jim Cane for preparing the maps, and to Gordon Davies for designing the jacket.

EXECUTIVE SUMMARY

The purpose of this study is to address some of the problems which stand in the way of gas development for internal markets in the Third World. Many countries in the Third World have proven gas reserves, and the geological prospects of further discoveries are held to be good. In some countries significant gas developments have already taken place, and several others are actively considering new gas projects. Yet even a cursory survey of the history of gas in the Third World immediately reveals that gas development for internal markets has been patchy, that it has lagged behind the emergence of new opportunities and that gas seems to suffer from a number of handicaps. These are evidenced either in an initial lack of interest in its development, or in considerable delays in agreement being reached between governments, companies and banks over particular projects, or in delays in the implementation of projects.

Before listing the main difficulties and some proposed solutions we would like to explain from the outset our general standpoint. Our study was not inspired by some missionary zeal in favour of gas; we do not believe that gas is special and do not share the views of those who, for political reasons, want to reduce *at any cost* the dependence of energy consumers on imported oil. Similarly, our study is not inspired by a general preference for oil, and therefore by an initial prejudice against gas. Our view is that gas is simply one natural resource amongst others, and that the fundamental economic principles which govern the appraisal of any natural resource project apply equally to gas.

Gas is only worth developing by a country endowed with reserves if the opportunity value of gas to the economy is higher than the cost of exploration, development, extraction and transport to the final market. The opportunity value of gas is set by the fuels which gas displaces: it relates, with some adjustments, to the price of oil at the country's border. However, it is the future, not the current, price of oil that matters for this appraisal. The conventional belief that the international price of oil will rise in the second half of the 1990s to well above current levels is a favourable factor in this context.

The costs of gas for a country include the opportunity cost of the domestic resources employed in the project (in its totality) and the compensation paid to foreign investors, lenders and contractors for their participation. Our position is that there is a prima-facie case for a government to consider a gas project only when a preliminary, but necessarily tentative, economic appraisal suggests that the future opportunity value of gas is likely to be perceptibly higher than the direct and indirect costs which the country will incur. We also believe that this simple economic appraisal, assuming oil prices at $15 per barrel in constant 1987 dollars, justifies the development of most *onshore* gas reserves in developing countries where gas is known to exist, and where an adequate domestic market can be promoted and developed, providing the distance between gas fields and markets is not excessive, and the terrain between them does not impose punitive transport costs. Some offshore reserves would also pass this test. More optimistic assumptions about future oil prices are needed to justify the development either of gas in remote locations or of offshore gas in deep waters.

To decide in principle that a gas reserve is worth developing is a preliminary (although major) step, but does not take us very far along the road leading, if all goes well, to the eventual implementation of a project. The obstacles and difficulties on that road are many. We attempt to identify the most important, and to suggest ways of overcoming them. Our recommendations derive from: (a) an analysis of the specific characteristics of gas and of gas projects; (b) a thorough assessment of the government/company relationship; and (c) the experience and information obtained from eight country studies and a large number of interviews.

Early problems relate to a priori conceptions about gas which can affect the attitudes of potential partners in its development (both in obvious and in more subtle ways):

(*a*) *One common misconception is to equate gas to oil.* True, gas is a substitute for oil products; true, gas upstream has many direct and indirect links with oil upstream and for this reason falls within the interests of oil companies. But the main similarities end there. The major differences between gas and oil are the absence of a developed, multilateral, international market for gas and the rather rigid link which must be established from the outset between gas reserves and dedicated markets.

(*b*) *Another misconception is to expect the premium physical qualities of gas to justify economic premiums in gas use.* The inherent qualities of gas may make it superior to oil or coal as a fuel from the physical and environmental points of view, but these qualities are only economically relevant if the government, the consumer or some other agent is prepared to pay for these added characteristics.

(*c*) Because gas is different from oil, there is a fundamental difference between gas projects for the internal market and oil projects, or indeed most other primary resource projects. The difference is that *the gas project must be appraised, planned and implemented as an integrated package involving upstream, transmission and markets.* A major difficulty arises as regards markets because they require initial promotion, rapid build-up to a minimum critical size and persistent efforts to ensure long-term development.

(*d*) Another difficulty is that *two different species of agents are involved in the upstream and downstream sectors of the gas industry.* Gas upstream is a mineral resource industry with the attendant problems of exploration risk. Gas downstream on the other hand is essentially a public utility, which provides an input to power generation and which also competes with electricity in the residential/commercial sector and in industry. The approach to upstream and downstream problems is inherently different; different agents are involved; and yet both downstream and upstream development are interdependent and must be planned and implemented together. Since these two agents are usually an international oil company and a government-owned agency the problem emerges as that of the relationship between a company and government. *Hence the relationship between governments and foreign companies occupies a central place in this book.*

Both parties encounter difficulties with their counterpart. Two major problems perceived by companies have led them to avoid involvement in gas projects for internal markets. Namely:

(*a*) It involves them in a greater dependence on and vulnerability to the state, over a larger number of issues and over a longer period of time, than is experienced with any other form of direct foreign investment.

(*b*) It will not generate direct foreign exchange earnings. Although benefits to the country are in the form of foreign exchange savings these are invisible. This poses problems for the foreign investor, such as an international oil company contemplating an involvement upstream, or for the foreign lender, particularly in countries known to suffer from a foreign exchange constraint or where the currency is not fully convertible.

Given the central problem of the relationship between companies and governments our main recommendations are designed to improve mutual trust which, needless to say, is a *sine qua non* for a successful venture:

(*a*) Companies need to recognize that governments pursue multiple objectives and are subject to diverse political pressures. An improved knowledge of the political economy of the country in question would help in avoiding unnecessary misunderstandings and irritations in negotiations, and would ensure that companies' initial requests and demands do not go well beyond what a government can possibly concede.

(*b*) Companies need experienced and sensitive negotiators for gas projects; and to revise their approach to the sharing of information with governments. The good negotiator knows that while withholding information may sometimes provide a tactical advantage in negotiations, sharing information and knowledge is likely to have strategic merit in laying the foundation of a trusting relationship.

(*c*) Governments need to recognize the importance of a unified position on the gas issue. The diversity of interests, and the large number of ministries and state agencies which need to be involved directly or indirectly in a gas development project cause problems both for the relationship with companies and the planning of the project. Hence, we suggest that governments may wish to appoint an individual or a special agency to the role of co-ordinator for the gas project. The functions of this person or agency would be to ensure co-operation of the various administrations involved with the project, to resolve conflicts of objectives, and to provide an effective channel of communication between them. We are *not* suggesting that the co-ordinator should take away responsibility from the relevant authorities, or usurp their roles in any way. The role of the co-ordinator would be strictly limited to that of facilitator.

(*d*) Governments can also help to promote mutual trust with companies. Abrupt and unpredictable changes in policies do much damage to trust, and bureaucrats can destroy the goodwill which higher echelons of government may be striving to build up. There is thus a strong case for avoiding sudden changes of course in matters of policy and for curbing, in some areas at least, the excessive zeal of bureaucracies.

(*e*) A gas project is therefore best conceived and implemented within a recognizable strategic framework. Both economic and energy planning are essential. The market for gas is interrelated to the market for all other fuels and to the economy in general, and since gas projects involve large investments they have significant macroeconomic implications for a developing country.

The government/company relationship finds its concrete expression in legal and contractual arrangements. The adoption of careful provisions in these arrangements can also facilitate this relationship:

(*a*) *There should be a specific gas clause in exploration contracts, distinct from the clauses on oil.* This clause should include a description of the initial steps to be taken if gas is found, and the principles of pricing and taxation which will be followed.

(*b*) *Producer prices should be market related, i.e. calculated in relation to the substitute fuel netted back to the well-head.* (Prices should not be used, either implicitly or explicitly, to claw back part of the rent: this function is better performed by fiscal instruments.)

(*c*) *The tax regime should be kept simple, progressive rather than proportional* (since proportional taxes are regressive) *and must allow companies a share of the upside potential of a project.* Sound principles of taxation should not be modified, as often happens, in order to provide additional incentives to companies; this causes unnecessary complication and costly distortions.

The development of a project, from the initial appraisal onwards should involve *joint* evaluation, followed by *joint* planning and *co-ordinated* implementation upstream and downstream.

(*a*) The initial step should consist of determining the possible commerciality of a find by carrying out a joint evaluation of the economics of the project as a whole, and the total benefit for country and company of developing the find. This implies a preliminary assessment of upstream and midstream investment as well as an assessment of potential markets.

(*b*) Once the commercial potential is agreed by the parties it is essential that dialogue and joint planning continue. All phases of the gas project have to be planned and developed in parallel and it is important to recall that all the phases will remain interdependent for the lifetime of the gas field. The preliminary commerciality decision is a major one, since it commits both parties to significant spending on the project; but the detailed field appraisals, costings and market studies need to be co-ordinated with each other, not carried out in isolation. For this reason it is helpful to establish a formal structure at the beginning of a project, defining the ways in which communication will be maintained. It will be necessary to define clearly those individuals with decision-making powers within each party, and to arrange for formal sharing of information and ideas at all stages. This structure should be designed to carry through to eventual implementation of the project and beyond.

(*c*) Market development will be one of the major areas requiring work. (Established gas utilities are able to provide valuable assistance in this area.) The cost of providing gas transmission and distribution infrastructure means that it is essential to find a bulk market for gas. The most likely candidate for this role will normally be the power sector; but electricity authorities frequently have no great sympathy for gas or experience in its use. They must therefore be brought into discussions from the outset. Early negotiations and early planning can be most helpful in overcoming the doubts and hesitations of major users. It will also be useful to identify at an early stage any premium markets which could be promoted in the country. While they will never be the sole justification for a gas development, due to their limited size in developing countries, they should not be forgotten.

Finally, we come to the problem of finance and to the role of international aid organizations. A gas project needs financing (a) upstream, (b) for the transmission system, and (c) in some cases downstream.

An international oil company involved upstream will provide the capital required for exploration, development and production, but will need guarantees on the repatriation of some of the principal and some profits. The nature of the required guarantee will vary from country to country depending on economic circumstances. Solutions involving payment in kind (in crude oil or products), or setting up a foreign exchange account in which part of the savings from oil imports are set aside, or some outside guarantee, all deserve the attention of negotiators.

Investment in the transmission system is a straightforward matter of project aid, and we believe that the World Bank and similar institutions have a privileged role in this area.

The financing of downstream projects must rely on conventional methods; in the final analysis the ability to raise funds for such a project depends on its economic quality.

The World Bank has an important role to play in all the phases of the development of a gas project.

(a) It can initiate and educate.

(b) It can be usefully involved as a neutral partner in the joint evaluation of a project from the preliminary appraisal onwards; and perhaps help by financing some part of the appraisal process. The involvement of the Bank in this process can reassure the partners and contribute to the building of mutual trust.

(c) In the final analysis the most effective ratification that the Bank can give to a good gas project is to provide a loan for the transmission infrastructure.

The recommendations summarized here but developed at greater length in this book do not constitute a panacea. They are essentially the results of analysis and debate, and constitute an independent assessment of what can be done. Nevertheless, they should be subjected further to the test of debate, and those that survive this first assessment should then be subjected to the harsher hurdle of trial implementation. We therefore propose a dynamic approach to the gas development issue, from informed analysis and policy suggestions, to debate and experimentation, rather than a book of easy recipes. This is simply because the solution to the gas development problem involves much more than altering this or that clause of a contract, this or that aspect of pricing or fiscal policies, however important these issues may be in themselves. Behind these lie problems of perceptions, attitudes and relationships between the parties concerned; and these can only be resolved through a painstaking effort of research, clarification and patient dialogue.

CONTENTS

TABLES

MAPS

ABBREVIATIONS

ADB	Asian Development Bank
AGIP	Azienda Generale Italiana Petroli
ASEAN	Association of South East Asian Nations
bbls	billion barrels
bcf	billion cubic feet
bcf/d	billion cubic feet per day
b/d	barrels per day
BMTA	Bangkok Mass Transit Authority
BNOC	British National Oil Corporation
BP	British Petroleum
btu	British thermal units
CNG	Compressed natural gas
DMR	Department of Mineral Resources (Thailand)
EEC	European Economic Community
EGAT	Electricity Generating Authority of Thailand
EGPC	Egyptian General Petroleum Corporation
ENI	Ente Nazionale Idrocarburi (Italy)
EPMI	Esso Production Malaysia Inc.
ETAP	Entreprise Tunisienne d'Activités Pétrolières
E£	Egyptian pounds
f.o.b.	Free on board
GDC	Gas Development Corporation
GdE	Gas del Estado (Argentina)
GDP	Gross domestic product
GNP	Gross national product
GRIP	Guaranteed Recovery of Investment Principal
GWh	Gigawatt hours
IADB	Inter-American Development Bank
IDA	International Development Association
IEDC	International Energy Development Corporation
IFC	International Finance Corporation
IGDC	International Gas Development Corporation
IMF	International Monetary Fund
INOC	Iraq National Oil Corporation

kg	kilograms
kg/he	kilograms per hectare
kgoe	kilograms of oil equivalent
km	kilometres
kt	kilotons
ktoe	kilotons of oil equivalent
KUFPEC	Kuwait Foreign Petroleum Exploration Company
LIBOR	London Inter-Bank Offered Rate
LNG	Liquefied natural gas
LPG	Liquefied petroleum gas
mb/d	million barrels per day
mbls	million barrels
mbtu	million British thermal units
mcf	million cubic feet
mcf/d	million cubic feet per day
MIGA	Multilateral Investment Guarantee Agency
MLNG	Malaysia LNG
mm^3	million cubic metres
mt	million tons
mtoe	million tons of oil equivalent
MW	Megawatts
n/a	not applicable
NEPA	National Electric Power Authority (Nigeria)
NESDB	National Economic and Social Development Board (Thailand)
NFC	National Fertilizer Corporation (Thailand)
NGLs	Natural gas liquids
NGOT	Natural Gas Organization of Thailand
NNOC	Nigerian National Oil Corporation
NNPC	Nigerian National Petroleum Corporation
NORAD	Norwegian Agency for Development
NP	Nitrogen, Phosphorus
NPK	Nitrogen, Phosphorus, Potash
OECD	Organisation for Economic Cooperation and Development
OECF	Overseas Economic Cooperation Fund (Japan)
OGDC	Oil and Gas Development Corporation (Pakistan)
ONGC	Oil and Natural Gas Corporation (India)
OPEC	Organization of the Petroleum Exporting Countries
OPIC	Overseas Private Investment Corporation (USA)
PCIAC	Petro-Canada International Assistance Corporation
PEMEX	Petróleos Mexicanos
POL	Pakistan Oilfields Limited

PPL	Pakistan Petroleum Limited
PRT	Petroleum Revenue Tax (UK)
PSA	Paul Sithi-Amnuai
PTT	Petroleum Authority of Thailand
SAFREP	Société Africaine d'Exploration Pétrolière
scf	standard cubic feet
SEGMA	Société d'Études de Gaz Marin
SGC	Southern Gas Company
SGTC	Sui Gas Transmission Company
SNGPL	Sui Northern Gas Pipeline Limited
SRT	State Railways of Thailand
STEG	Société Tunisienne d l'Électricité et du Gaz
tcf	trillion cubic feet
t/d	tons per day
toe	tons of oil equivalent
TORC	Thai Oil Refinery Company
TP	Texas Pacific
TPDC	Tanzania Petroleum Development Corporation
t/yr	tons per year
UNCTC	United Nations Centre on Transnational Corporations
USAID	United States Agency for International Development
YCF	Yacimientos Carboníferos Fiscales
YPF	Yacimientos Petrolíferos Fiscales

PART I

ISSUES AND POLICIES

1 GAS IN DEVELOPING COUNTRIES: RESERVES, COMPETING FUELS, MARKETS AND THE BEHAVIOUR OF ENERGY PRICES

1.1 Introduction

The purpose of this chapter is to address a number of preliminary, but essential questions about natural gas development in the Third World. We also identify and then challenge a number of preconceptions about gas development, which can sometimes hinder the progress of gas projects.

A necessary, though by no means sufficient, condition for gas development is the availability of adequate reserves. Where reserves have not yet been found, good discovery prospects could attract explorers and set the first stage of gas development into motion. We shall therefore begin (Section 1.2) with a succinct appraisal of the state of gas reserves in the Third World.

Gas is not a unique fuel. It has a wide variety of both energy and non-energy uses, but in each possible use there are one or several substitutes to gas. It is thus important to compare the physical and economic characteristics of gas with those of other fuels and energy sources (Section 1.3).

Fuels are not necessarily substitutes for each other across the board however. They normally compete against each other in particular markets. Furthermore, gas development is only possible when production from discovered reserves finds market outlets and stands a chance (a) of displacing competing fuels and/or (b) of securing a share of incremental demand for energy. We shall therefore address the issue of potential markets for gas in Section 1.4.

Finally, the important economic variable governing both the profitability of investment upstream and the rate of gas penetration in markets is its price. We shall have much to say on many aspects of the gas pricing issue in other chapters. In Section 1.5 we shall concentrate however on a preliminary question of broad significance: are the expected future movements in oil prices a deterrent or an incentive to gas development in today's economic environment?

1.2 Gas Reserves

A large number of developing countries have gas reserves. While the bulk of proved gas reserves in the world (approximately 55 per cent) lie in the North (USSR and OECD countries), the remainder is to be found in the South. Some of these countries are relatively richly endowed with gas, with a further 35 per cent of reserves being found in just ten countries, each with at least 45 tcf of gas. (Iran, Qatar, Algeria, Abu Dhabi, Saudi Arabia, Mexico, Venezuela, Indonesia, Malaysia and Nigeria – in decreasing size of reserves.) But the remaining 10 per cent of proven reserves are distributed through a large number of countries in the South. Many have quite small reserves compared with the world total; yet these reserves are often large in relation to the potential of the internal markets. Exploration even for oil in these countries has been rather sporadic, and gas as such has rarely been looked for. The prospect of further gas being discovered in significant amounts in the Third World therefore seems quite high. In fact, it is likely that in future more gas than oil will be discovered in the world at large, and that the ultimate size of remaining world gas reserves is greater than those of oil.

Table 1.1 lists twenty-nine Third World countries with known gas reserves, and whose potential internal markets are not negligibly small. This list suggests that the opportunities for gas development, in principle at least, are both diverse and widespread through Asia, Latin America and North Africa.

Table 1.1: Third World Countries with Gas Reserves.

Asia	Africa	Latin America
Abu Dhabi	Algeria	Argentina
Afghanistan	Egypt	Bolivia
Bangladesh	Nigeria	Brazil
Burma	Tanzania	Chile
China	Tunisia	Colombia
India		Ecuador
Indonesia		Mexico
Iran		Peru
Iraq		Venezuela
Malaysia		
Pakistan		
Qatar		
Saudi Arabia		
Syria		
Thailand		

An important distinction must be drawn at the outset between associated and non-associated gas. Since associated gas is a by-product of oil extraction, the problem of its use arises as soon as crude oil is produced from an oilfield. In the early days gas was not seen as especially valuable: it was sometimes reinjected but often treated as a waste product, and flared. Through the 1960s government and public opinion gradually began to perceive such flaring as wasteful, and legislative and economic measures were introduced to curb flaring, either through reinjection of the gas or through finding a use for it. Flaring does still exist, but is now generally seen as an anomaly, almost as a scandal. In the 1970s the opportunity cost of associated gas rose dramatically along with the price of oil, accentuating the political pressure to make use of the gas. Each btu of gas used within an oil-producing country can release a btu of oil for export, either immediately or in the future, or can save a btu of imported oil. A few countries with huge oil reserves are of course less affected by this argument; but countries with smaller reserves such as Egypt, Tunisia, Nigeria or Pakistan can gain immediate benefit from substituting gas on the domestic market, and releasing more oil to earn (or save) foreign currency.

The availability of associated gas in oil-producing countries is thus, in a sense, a spur for promoting the utilization of this resource. Yet in many countries much still remains to be done to achieve optimal use of associated gas. Nigeria, one of the countries covered in this study, is a case in point. Of course there are sometimes problems with associated gas. Production rates are not always very dependable because they are determined by the autonomous factors that govern crude oil extraction. This makes it difficult to justify capital expenditure on a gathering system, pipelines and gas-burning equipment – all of which require a constant and dependable supply of gas. A further problem arises over the pricing of associated gas. A common view amongst governments is that since associated gas is a by-product of oil extraction it is of little value to the producing company (a view reinforced by the long history of flaring). Governments are thus reluctant to pay producers the same price for associated as for non-associated gas. This may weaken the incentive for oil exploration and deprive the country of further discoveries of both oil and gas. The government position has a valid rationale; and yet it is important not to lose possible incentives to explore and develop. One solution is to ensure that the package of incentives for hydrocarbon exploration compensates for the low price offered for associated gas by providing more generous terms on other items of the agreement.

Non-associated gas is often discovered when companies are exploring for oil. There are of course exceptions, where exploration is aimed specifically at finding gas; but in the Third World these cases were extremely rare in the past and are only slightly more frequent today. Non-associated gas has certain advantages over associated gas. For example if suitable markets do not currently exist development can be postponed and the resource will not be wasted through flaring. When production does take place the extraction rate is not governed by an exogenous factor, as for associated gas. Furthermore the issues relating to possible development of a reserve are somewhat clearer; they are at least free from the additional complications of joint production of oil and gas.

1.3 Gas and its Substitutes

Since gas is just one energy source among many, it is important to look closely at the properties of gas in relation to its substitutes. This will define more clearly the parameters of its potential. Gas is often claimed to be a premium fuel, because it burns relatively cleanly and is very flexible. In particular it is possible to concentrate and direct the flame more easily than with other fuels. These premium qualities do give a definite advantage to gas in certain applications, such as the glass and ceramics industries. Its non-polluting characteristics can also be important in urban environments where pollution has become a pressing problem (e.g. Athens, Ankara, Mexico City to name but a few). But these advantages of gas are not enough in themselves to ensure its development in most Third World countries with limited industrialization. In order to establish a bulk market, which can justify the capital expenditure on field development and infrastructure, gas needs to compete in applications where alternative fuels such as oil, coal and hydro are just as suitable.

This is the paradox of gas. One school of thought argues that great weight should be given to the physical or chemical properties of any resource – in other words the premium qualities of gas. According to this school, burning gas under a boiler is equivalent to making saucepans out of gold, and selling them at the same price as aluminium pans, simply because one has no other immediate use for the gold. If there is no high-value market immediately available for gas then it should be preserved in the ground, until such time as its potential value can be realized. We believe that this is a misconception, because the economic value of a commodity does not depend on its physical properties alone. The economic factors that govern its present and future use cannot be ignored. Gas development today depends on capturing bulk markets, and gas is therefore inevitably in competition with other low-cost fuels. It is difficult to envisage a situation in the Third World, even in the future, where the sequence of gas development will not begin with the penetration of a bulk (non-premium) market which may lead in time to the growth of premium uses. It is true that in Europe and the USA a different sequence obtained, but for historical reasons which will not recur. In the nineteenth century town gas was introduced in these continents because there was a demand for efficient lighting and consumers were prepared to pay a high price for this highly valued commodity. This led to the installation of a distribution network for residential and commercial purposes (the premium sector) which enabled a continuation of supplies when the use pattern switched away from lighting. When natural gas was later introduced the infrastructure was already in place and natural gas was able to penetrate immediately the premium sector. Nevertheless, while it is important to emphasize that gas development in the Third World must involve bulk use, the insight of the 'premium' thinkers should not be lost altogether. Gas does retain certain qualities which differentiate it from oil and other fuels even if it cannot always capitalize on them; perhaps this should be used as a reminder, in countries where reserves are not very large, to limit the quantity that is burned (or even flared) in low-value uses.

Every fuel has its own properties which give it advantages for certain markets or uses. The properties of each competing fuel as compared to gas therefore need to be assessed. A brief summary is given below.

(a) *Gas* v *Oil*. In most of the potential markets for gas, energy requirements are currently met by utilization of an oil product. In a broad sense, oil and gas are therefore major substitutes. To think of them in exactly the same terms however is very misleading, and may even have caused some of the problems in developing gas as a fuel. There are in fact important differences between the two fuel.

The first of these differences lies in the *markets* for oil and gas respectively. If a country discovers oil reserves it has a commodity which it can easily sell. There are existing markets for oil in all countries of the world, and well-established uses for oil products. Over a wide range of market sectors the normal fuel is oil and the international market is sufficiently competitive to absorb with ease newcomers with small additions. In contrast a gas field can only be developed for a specific market outlet. Even if the field is large enough to have export potential it is essential that a market be identified, and firm sales arrangements be made with a customer in advance of field development. The contrast between finding small oil reserves in

Angola, and developing the Indonesian gas reserves for LNG exports to Japan demonstrates this difference clearly.

Because gas development involves dedicated markets, and a rigid transportation link between the well and the final outlet, a gas project must be conceived as an interrelated package including upstream investment, transmission, and the promotion of markets downstream. Furthermore, the development of these three parts of the project needs to be carefully planned and synchronized (see Chapter 4). By contrast, an oil project does not usually involve the development of specific market outlets and can thus be restricted to oilfield development and to transport facilities between the field and a refinery or an export terminal.

The marketing of gas is not straightforward. For internal markets it is necessary to promote gas as a substitute for existing fuels. Customers will need to be persuaded of the advantage in conversion, both because gas will be unfamiliar, and because capital costs will be involved in altering equipment. This means that a price incentive will often be required. Customers will also want to be reassured that such a price incentive will be long-lived, and not just a temporary phenomenon. A further problem arises with gas because the promotion of the market and the confirmation of reserves must be synchronized when planning a project. Companies are reluctant to spend money on further exploration for gas until firm market outlets have been identified. But promoting a market requires the availability of gas to serve it.

A second major difference between the two fuels lies in the *transport* of gas and oil. Oil is consumed as products, rather than as crude. The immediate need is therefore for a pipeline or a truck to take oil from the field to a refinery or an export terminal. From there onwards it can be distributed in pipelines, tankers, trucks, or barges. There is a great choice both of methods of transportation and size of cargoes. The products can be easily dispersed over innumerable markets worldwide. In contrast gas is taken from the field to its final point of use by pipeline. While it is technically possible to transport small quantities by truck, this involves the need for compression or liquefaction and containers, making it relatively less advantageous to transport small quantities in this way. In addition the gathering costs are higher for gas than for oil, and gas transmission pipelines are more expensive per btu than oil pipelines. This discrepancy in cost becomes more marked for larger volumes of gas over long distances. (All such comparisons are of course influenced by the costs involved in developing specific fields, and by prices and tax regimes.) Even when exporting very large quantities of either fuel, oil has a cost advantage: oil can be fed straight from a pipeline into tankers which hold vast quantities of liquid, are structurally very simple and therefore relatively cheap; for gas, expensive liquefaction and regasification facilities are needed at either end of an LNG transport link, in addition to costly dedicated tankers.

The *relation* between *producer* and *buyer* is very different for oil and gas. The technical characteristics of oil enable a producer to deal with a multitude of buyers, if required, whereas a gas producer typically deals with only one, or at most a few, buyers. This leads to very different economic relationships, and a very different market structure. The gas market is characterized by bilateral monopolies/monopsonies, compared to the atomistic market for oil. Furthermore, the relationship between gas producer and buyer is inherently long term because of the rigid links between the production site and dedicated markets.

Where the producer is an international oil company,[1] its relationship with the government is important for oil production, to ensure amicable agreements over depletion rates, taxation, employment practices, and so on. But in the case of gas the relationship covers a much wider range of issues. This is especially the case if gas is destined for the internal market, since the government or a parastatal organization is the usual customer for the gas. The most obvious additional interface between the two parties is over the price to be paid for the gas, since there is no obvious international reference price as in the case of oil. An extra dimension is also added to the relationship by its long-term character. A company not only needs to reach agreement with the government over current prices, but needs to be confident that future governments will also be amenable to negotiation (in response to unanticipated changes in circumstances) and be willing to concede new terms, if necessary, to protect the cash flow of the project. These issues are discussed at length in Chapters 2, 3 and 4.

A fourth difference between oil and gas lies in the *cost-benefit aspects* of projects. As a rule, cost-benefit appraisals will be carried out by both governments and companies, but from different standpoints. The government has a duty to consider a number of externalities, whereas the company will only do so if this seems expedient in a particular case. (This point is developed in Chapter 2.) This initial difference in attitude is exaggerated in the case of a gas (as opposed to

---

[1] Throughout this report 'international oil companies' refers to foreign oil companies which operate (or are able to operate) in a number of countries, whether they are privately owned (e.g. Shell or Exxon) or owned by the state (e.g. AGIP or Petrobrás). The term 'international' refers to the geographical scope of their activities and not to ownership.

oil) project, because the external effects of a gas project are generally more diverse and widespread than in the case of oil.

From the company's point of view physical differences in the production of oil and gas have implications for project appraisal. The recovery factor from a gas field is often as high as 70 or 80 per cent. The production facilities required to recover this gas from the field (unless the field is very large when development may be in stages) are usually installed in one lump at the beginning. By contrast, the recovery factor from an oilfield is low, the typical range being 15–30 per cent. In many cases further investments in production facilities are made at later stages in the life of a field to increase the recovery factor. The economic appraisal of a gas project will therefore typically have to take into account the total investment for production, made in one lump at the beginning of the project;[2] whereas an oil project can be appraised on the basis of just the initial investment, leaving future additional investments for a separate appraisal, to be made nearer the time of the future undertaking (especially if they will not be required for some time).

Thus an oil project may be split into two or more separate consecutive projects for the purpose of economic appraisal, while a gas project will normally be appraised as a single undertaking with most of the capital investment up front. Under these conditions the appraisal method puts gas projects at a disadvantage *vis-à-vis* similar oil projects because the income stream of the gas project accrues over a longer period of time, and the income generated in the distant future adds very little to discounted present values.

Lastly, the *attitude of oil companies* towards gas is different from their attitude towards oil. In today's world it is most likely to be the oil companies which find gas. As already mentioned some gas is associated with oil, and thus produced from the same wells. Non-associated gas is sometimes found in the same areas as oil, and thus may be found when a company is exploring for oil. And even if exploration were specifically for gas in areas known to be gas prone, similar techniques would be required, and these are the province of the oil companies. Oil and gas are therefore dealt with upstream by the same companies, whether they be foreign or national oil companies. This apparent link between the two fuels introduces however important elements to the competition between oil and gas as fuels, beyond the economics of their use in various markets. For the competition between the fuels becomes to some extent internalized within the oil

---

[2] Of course a gas project may entail additions at a later stage, such as an extra compressor. But these are small investments compared with those made for secondary or tertiary recovery for oil.

companies. Corporate objectives and strategies with regard to the fuels become crucial. For example a company may decide not to encourage development of gas finds, if the opportunity cost for this would be a reduction of profit on its oil business. (A form of restrictive practice.) These instances are probably rare because of competitive threats from other firms. But nevertheless it is important to remember that gas projects compete for scarce funds and skilled manpower with other investment projects available to a company. The decision to undertake a gas project not only depends on the absolute expected profitability of the venture but on its profitability (and other relevant features) *relative* to other investment opportunities.[3] From a different perspective a company may decide to promote gas development in a market where this would not be justified on grounds of immediate benefits, in order to gain a foothold in a new country or a new market for its oil activities. Again, a gas development for the internal market may be more attractive to a company which already produces oil in that country than to a company with no oil production there; the former may obtain oil for export in exchange for the gas sold domestically, and would thus have fewer worries about foreign exchange convertibility.

A further point is that oil companies have different amounts of expertise in oil and gas. Companies typically have the ability to become involved in all stages of the oil industry, both upstream and downstream if they so choose. They therefore have the capacity to take full control over an oil project. Downstream gas activities on the other hand have very rarely been part of their business. While some companies express willingness to become involved in building gas pipelines and marketing gas this lies outside their traditional areas of competence.

The attitudes of oil companies towards gas will be discussed at length in Chapter 2. Suffice it to say at this point, that the competition between gas and oil is fundamentally affected by the interests and investment strategies of the companies, and is by no means limited to a straightforward comparison between the advantages of each fuel.

(*b*) *Gas v Coal.* The market for coal is intrinsically more limited than that for either oil or gas. Gas comes into competition with coal where it is used for steam raising in bulk market applications, such as power generation and industries. In analysing the problem of gas substitution for coal two main cases must be distinguished. First, the displace-

---

[3] In this respect gas is often at a disadvantage because there are generally greater opportunities available to companies for profitable oil projects. Of course small oil projects in countries with no export potential suffer in the same way as gas.

ment of coal by gas in existing coal use, and secondly the choice between gas and coal for a new power or industrial project, particularly where the transport (and handling) infrastructure for neither fuel pre-exists. In the first case gas will only displace coal if it is priced competitively at *the burner-tip*. Gas has no premium qualities to offer in steam-raising applications, but has the advantage of lower handling costs *in situ*, flexibility/simplicity in use, and, of course, environmental merits. Supply flexibility is however greater for coal than for gas. The competitive price of gas at plant gate may thus have to be close to that of coal, as some of its disadvantages cancel its merits (other than environmental, which if valued by society call for a subsidy from the state).

In the second case, where a completely new project is under consideration, the choice between gas and coal raises more complex issues. Coal infrastructure for transport and for handling at both the harbour (if coal is imported) and the point of use involves large capital expenditure and the provision of large tracts of land for storage. Gas is of course expensive to transport, but there are virtually no handling costs as such. A further complication arises from the difficulty of assessing realistic long-term coal prices. Under today's circumstances it is possible to buy coal at very low prices; but such prices are unlikely to be relevant when developing new projects, as they are a manifestation of today's glutted markets. Prices which would justify the development of new coal-mines, or at least expansion in capacity of existing mines, would be more relevant. There is, unfortunately, considerable divergence in opinion over how such prices will move over the next decades. Some experts believe that they will be twice or even three times higher than prices for marginal quantities, whilst others feel that there will always be adequate availability of very cheap coal. For meaningful comparison, one might have to study in detail where coal for a new project might come from, and how the price from that destination might behave. Generalizations on the relative economic merits of gas and coal are unlikely to be possible when it comes to considering new user capacity.

(c) *Gas* v *Hydro*. Most countries in the world have some potential for using hydropower to generate electricity, although this will of course depend on suitable sites being located close to markets. The great advantage of hydroelectricity is that it is in a sense a renewable resource, though the silting up of dams is an important limitation. It also requires very large initial capital outlay, which can lead to serious financial difficulties. This is only partly allayed by the fact that a hydro dam is a highly visible and often visually attractive showpiece, making the funding of hydro projects relatively popular with aid agencies. There are other disadvantages of harnessing hydropower, which are often rather difficult to quantify, but are beginning to be more widely recognized. In particular there is a problem over competing uses for water, especially in tropical developing countries, where maintaining reservoir height may mean depriving agriculture of water supplies. If agricultural requirements are given precedence the power supply becomes undependable. Furthermore the stability of electricity supplies comes to rely on regular rainfall, which may be uncertain. Building dams also involves a loss of agricultural land in valleys, and often requires the resettlement of populations. There are also health issues associated in some cases with the creation of lakes behind the dam. There are therefore definite limits to the usefulness of hydro schemes; and very few countries can afford to rely solely on this source of electricity. One would not however expect gas to completely displace hydro. The scope for such substitution is discussed in Section 1.4.

(*d*) *Gas* v *Nuclear*. In most developing countries there is very limited competition with gas from nuclear power. Nuclear is at a disadvantage because the minimum economic plant size is so large that it would often imbalance the power system. Furthermore, a combination of high capital costs and external political pressures makes it unlikely that nuclear power will play a large part in the energy balance of developing countries for the foreseeable future, except in the largest countries – China, India and Pakistan. One must recognize however that certain countries with important gas reserves still express the desire to build nuclear power stations, perhaps for political reasons.

(*e*) *Gas* v *Biomass*. The severe problem caused by depletion of biomass resources in many developing countries raises the question of whether gas could usefully substitute for biomass. But the underlying problem is that biomass consumption is highest in rural areas, where relatively small absolute amounts of fuel are used over very wide areas. A gas pipeline infrastructure is prohibitively expensive in this context. The cost of transporting gas, even as CNG, tends to militate against its use. A more economical, because more easily transported, substitute would normally be kerosine. For this reason we do not see much prospect for natural gas as a substitute for firewood, even if the social cost involved in burning biomass is fully accounted for. LPG and kerosine are the most likely substitutes.

1.4 Markets for Gas

When gas is introduced into the energy mix of a country it will usually compete with different fuels in the different market sectors. Moreover, even if the dominant form of energy used in all sectors is oil (as may well be the case), gas will fare differently in the competition in each sector. It is therefore useful to look at each of the major energy-using sectors individually, to assess the potential for natural gas, and the specific problems which it may encounter.

The primary problem when faced with a new gas resource is to find a sufficiently lucrative market to cover the very large capital investment in pipelines that is required to bring the gas to its market. This means that either a market must be found in which the premium qualities of gas can command a very high price; or a very large market must be identified, where the quantity of gas consumed will compensate for its relatively low price, and thus still cover the capital investment. Even in industrialized countries, save in places where town gas had already created a market and provided an infrastructure, it has been hard to find a premium market (whether in industry or the residential sector) large enough to justify building a gas distribution network. Many countries have needed to put gas first into a bulk market, albeit many have succeeded in moving away from low-value uses over time. In the case of developing countries, often with no existing gas infrastructure at all, there is an overriding requirement to find a bulk market for gas. Once the infrastructure is in place, and costs are covered by reliable customers, there is certainly room for promoting gas use amongst smaller commercial and industrial consumers, who may be in a position to pay a premium for the fuel. Over the lifetime of the pipeline, and as the country's economy develops further, such premium markets may take an increasing share of the gas production. It is arguable that as this occurs the original bulk users should be transferred away from gas use. The precise modalities of such developments will however vary considerably between countries, depending on their overall level of economic development, the nature of their energy markets, and on the alternative fuels available at given prices.

These points are a reminder of the need for a long time-horizon in planning gas development. They also point to an important link between gas planning and more general economic development plans for major gas-using sectors such as power and industry.

(a) Power. Gas development depends critically on the existence of a bulk market with a dependable demand for fuel. The power sector is

the major candidate in most developing countries. All developing countries have a large demand for electric power, and in most this demand is growing at a very fast rate, often faster than GDP. There is thus potential for using gas in the power stations which will be built to meet the increase in demand as well as room for conversion of existing capacity. Furthermore serving the power sector provides opportunities for capturing other markets. A pipeline to a large power station will also usually take the gas to its other potential markets, since the major demand for electricity is in large conurbations, where most industry is also situated, and the residential and commercial markets are also to be found. In rare cases it is more economic to build a power station close to a gas field and transport the electricity rather than the gas, because of the terrain or other factors. But the potential advantages of bringing gas to a variety of markets, by transporting it to power stations close to, or on the route to, industrial centres or large towns, will usually outweigh this.

The case for using gas in the power sector rests on a number of factors. To begin with there is an environmental case when coal would otherwise be used, given the pollution problems in many cities in developing countries. A different but most important argument in favour of gas is that it produces low-cost electricity efficiently in combined-cycle power systems, now that the early difficulties encountered with this technology have been overcome. Furthermore, the capital costs of combined-cycle are low. These arguments, however, should not distract one from the important point that the use of gas in the power sector depends critically on the cost of supplying gas. For in this market gas is competing against fuel oil, coal and hydro. Furthermore, except in the case of hydroelectricity, this competition occurs not only at the time of the initial investment decision, but throughout the life of the power station. For a station converted to run on gas will normally also retain its original capability; and many new stations also have dual-fuel capacity. Should the relative prices of fuels change, it is often a very easy move for the power authority to switch its allegiances. This has been very clearly illustrated recently in Thailand, where a drop in fuel oil prices was not at first matched by a drop in gas prices. The Electricity Generating Authority of Thailand (EGAT) was understandably very anxious to revert to burning fuel oil, thus saving considerable amounts of money. The only solution to this situation was to bring down the price of gas, since the government agency, PTT, which sells the gas to EGAT was committed to a take-or-pay agreement for the gas. Thus the price of gas must be set in close competition with that of coal and fuel oil (whichever is the lowest-priced substitute), taking fully into account their respective merits and disadvantages. In competition with hydropower, the cost factors are less easily quantified, as discussed above.

Provision of electric power is not a simple matter of choosing between one fuel and another however. A balance needs to be struck between hydro and thermal (where hydro is available), and within the thermal sector between alternative fuels. The varying demand pattern, with peaks at various times of day and in the different seasons calls for a flexible supply system. Typically a country will need a certain number of installations to supply the base load of demand, which remains fairly constant; this will be supplemented by flexible stations which generate electricity only as it is needed in peak periods. This dichotomy is further complicated by the existence of hydro-electric plants, since their capacity is also usually seasonal in developing countries. During the wet season they will be optimally used to their full capacity, to provide a base load; but during the dry season their capacity will be much reduced, and will need to be supplemented elsewhere in the system. The problem of seasonality of hydro capacity cannot really be solved through the use of gas however. The major reason for using gas in the power sector is because of the large and constant demand for fuel, and a purely seasonal offtake of gas for power stations would defeat the purpose of covering capital infrastructure costs.

It is therefore necessary to look at the particular properties of gas in this perspective. To use the power sector as a bulk market for gas assumes that it will be used to provide a base load facility. This is how a steady and large demand for the fuel can be guaranteed. But this proves not to be always the field in which gas wins out against its competitors. Gas turbines, for example, are relatively cheap to build, but have a low efficiency, of about 25–30 per cent. They tend to be economic only when gas prices are very low, or they are run at low loads to meet peak demands. Such cheap additions to generating capacity can be extremely useful, especially in situations where capital is scarce; but they do not meet the requirement of justifying in itself the development of a gas field. Gas-fired steam turbines have a higher efficiency, of 30–35 per cent, but are expensive to build. They are run at high loads, to meet base demand: but here they are in direct competition with both coal and fuel oil, so the gas price must still be kept rather low. The only advantage of gas over coal and fuel oil in such boilers, is that operating and maintenance costs are lower for gas. Where gas begins to show clear advantages over other fuels is when it is used in specially-built combined-cycle power stations. In these the waste heat from a gas turbine is used to raise steam, and the overall efficiency is very high, at about 45 per cent. As mentioned earlier, the capital cost of such installations for gas is also quite low, making them a strong candidate for new additions to generating capacity.

Given this background it is interesting to look at how far gas has in fact penetrated the power sector. Most developing countries which are already consuming gas have focused on this sector as the major bulk user. Thus in recent years 98 per cent of gas consumed in both Tunisia and Thailand went to power stations; 58 per cent in Egypt, 31 per cent in Pakistan and 25 per cent in Argentina. A survey of government intentions for additional capacity over the period 1985–2000 has shown that most do not give great prominence to gas, in terms of the total generating capacity. In Latin America 55 per cent of additional capacity will be hydro-generated; and the comparable figures for Africa are 22 per cent, and Asia 32 per cent. In contrast planned expansion of gas combined-cycle capacity will provide 5 per cent for Latin America, 2 per cent for Asia, and 5 per cent for Africa. The total consumption of gas associated with these plants will be approximately 500 bcf per annum. Thus gas is not being heralded as the fuel of the future for power generation.

The reasons for the relative neglect of gas in electricity planning rest on a collection of received ideas and conventional wisdom, which have worked to obscure the true scope for gas in this sector. These include a simple lack of knowledge about the different technologies available for gas use and their relative costs, especially combined-cycle; a natural conservatism which works against the introduction of unfamiliar fuels, and which seems to particularly affect the authorities in charge of power, often known to strongly resist, or show lack of interest in, the adoption of gas; and long time-lags between the acceptance of new ideas in theory and their integration into economic plans. To add to these problems is the factor mentioned earlier, that aid finance is often more easily come by for prestigious hydroelectric dams, than it is for more mundane gas pipelines. Last, but not least, governments will hesitate to plan future electricity supplies on the basis of uncertain gas reserves. There is thus a need for considerable exploration and appraisal of gas potential before long-term plans can be expected to favour gas use.

These problems will need to be overcome if gas is to carve out a niche for itself; but it is nevertheless true that even if a large proportion of gas production goes to the power sector gas is unlikely to provide a *large* share of total power generation. Although in any given country the power sector may be a bulk customer for gas, large amounts of electricity will also be generated from other sources. Few countries have large enough gas reserves, exploitable at low enough costs, to displace competing fuels on a very large scale. Gas should be used in the power sector more widely than it is at present, but through careful planning. The most common situation will be where use in a power station provides the economic rationale for initial development of a gas field and construction of a pipeline: with the long-term aim of moving into the industrial and commercial energy sectors. It is not necessary for gas to capture a large *share* of the power sector in total, so long as the portion it takes represents a large *volume*, as this will be vital for the use of gas in other sectors as well.

(b) *Industry.* The market for gas (as a fuel) in industry divides into two distinct categories. On the one hand there are bulk markets, in which large amounts of fuel are consumed, but no special qualities are required other than the generation of heat. On the other hand there are more specialist uses, in which gas has a premium value.

Bulk users of fuel include aluminium and iron and steel smelters. But the commonest bulk market for gas in industry is in cement production, since most developing countries have their own cement plants. A number of countries with cheap gas have already substituted it in this industry. For example the Siam Cement Company in Thailand was one of the earliest customers for PTT's gas. In Egypt too 15 bcf per annum is consumed by this sector (8.5 per cent of total gas production, compared to 6.7 per cent by all other industries). The problem is that both fuel oil and coal are equally good fuels for this purpose; coal in fact has a positive advantage, in that the ash produced when it is burned can itself be used as a raw material in the production process. The economic value of burning gas in a cement plant therefore tends to be somewhat marginal when prices of fuel oil and coal are low. Nevertheless some countries such as Tanzania, which is very short on foreign currency, and where the one cement plant alone accounts for 10 per cent of all fuel oil consumed in the country, see a distinct advantage in using indigenous gas rather than exportable fuel oil. It also needs to be remembered that setting up the infrastructure for importing coal is itself a very costly process, even if the fuel itself is cheap. A study is currently under way of the relative advantages of using imported coal or indigenous gas for the cement industry in Tunisia, and the same issue is also being addressed in other countries.

Industries where gas has a premium value include ceramics, glass and textiles. The sulphur content of oil makes it unsuitable for ceramics production; and the close control over the intensity and direction of gas flames is of great utility in both ceramics and glass. In textiles, gas is well suited for drying which is an important part of the dyeing process. Where gas is piped to an area with these industries it will usually be worth their while to convert to its use. They do not in themselves consume enough energy to justify the special construction of a gas supply however.

The same stricture applies to a very wide range of other industries, such as brewing, food processing and light engineering industries, which are to be found in most developing countries. They do not consume specially large amounts of energy; and in addition they do not gain any particular benefit from using gas rather than alternative fuels (including electricity). If a gas supply is available however, perhaps because of a pipeline bringing gas to a nearby power station, and if the gas is competitively priced, then they can cumulatively provide quite a large additional market. From the consumer's point of view gas will always have the advantage that there is no need to provide any storage facilities for the fuel, and the tap can be turned on and off as required. From the community's point of view there will be significantly less air pollution if the majority of industries burn gas rather than coal or fuel oil. Both these advantages are difficult to quantify however.

(c) *Residential and Commercial.* The residential and particularly the commercial sector is usually quoted as the premium market *par excellence* for gas. For applications such as cooking many consumers prefer gas even to electricity, thanks to its superior controllability. For space heating too it is a very convenient solution: it needs no storage facilities, is clean, and provides fast heat. Compared to the burning of coal in households gas is also virtually pollution free; a significant improvement in the environment can result from moving consumers from coal to gas. However coal is not used in this sector in many developing countries, so the decreased pollution from switching to gas in this sector may not be very considerable. Residential consumers in developing countries are aware of some of the benefits of gas, and several governments have made useful political capital from providing domestic consumers with gas. This political effect is heightened, because it is normally only the wealthier classes that use commercial fuel in their homes, and these are the groups in society with political influence and weight. This effect has been especially marked in Argentina.

The overall potential for gas use in the residential market is said to be very limited in the developing world. The major reason adduced is that household use is usually limited to cooking, which uses only minute amounts of gas. But in certain cases there may be wider opportunities, and if a distribution network is installed care should be taken to capture all such potential. For example, households in Cairo were found to use the equivalent of one-and-a-half bottles of LPG per month for cooking, and the gas tariff structure was designed to discriminate against use in excess of this amount, on the assumption that higher-income groups could pay a higher tariff for additional use. Even low-income families in Cairo however it is common to use fuel to heat water for laundry. If the tariff structure were to take this into account the total demand of the residential sector would increase substantially, thus lowering the unit cost of supplying the gas.

Three factors may improve the market outlook. First, if the country has a fairly cold winter, producing a need for space heating, then consumption will of course be significantly higher. This is the case in both Argentina and northern Pakistan. Secondly, commercial premises consume much larger amounts of gas than private households. For example it has been estimated in Bangladesh that a restaurant uses the same amount of gas as twenty-seven households and a bakery as ninety households. A large first-class hotel, with water heating and air-conditioning may use as much as 800 households. If cities have a concentration of hotels and small commercial premises, and a large number of households with incomes sufficiently high to generate a demand for water heating, space heating or air-conditioning, then there may well be an economic case for building a gas distribution network to serve them. This has been recommended for example for some of the Mediterranean tourist resorts in Tunisia. Thirdly, where new suburbs are being constructed it is relatively cheap to supply gas, especially to high-rise buildings. This could somewhat improve the prospects for gas in the residential sector.

In normal circumstances however there is little economic justification in providing domestic gas for the majority of the population in developing countries, except in very large cities or in the major towns of cold countries. Both LPG and kerosine are much cheaper to distribute. Once one looks beyond cities to the countryside this argument is even stronger, for the population tends to be very widely dispersed. While there is a recognized problem of depletion of biomass resources, this can be solved better through the use of kerosine and perhaps LPG than with gas. It should also be pointed out that in countries where the natural gas is rich then gas production will directly benefit this market sector. If LPG can first be separated from gas, it can reduce the need for imports, and solve shortage problems which are currently exacerbating the depletion of firewood.

(*d*) *Methanol and Fertilizers*. The use of gas as feedstock to produce tradable commodities has seemed very attractive to developing countries. Such transformation has traditionally been seen as a 'higher-value' or premium use for the gas, as compared to burning it in power stations. For such processes are making use of the intrinsic qualities of the gas, rather than its ability to produce heat, which it shares with oil and coal. The second major attraction is of course the fact that products like ammonia or methanol are internationally tradable, and can earn hard currency.

The premiss for these arguments is that the product made from gas can earn a good price on international markets (or can substitute on the internal market for expensively imported products). At times this has certainly been true. But unfortunately the international markets for both fertilizer and methanol are not very encouraging at present; and there is little prospect of them recovering in the medium term. There has been great expansion in the last decade by new, very efficient suppliers, using exceptionally cheap gas resources. Against this background there are few developing countries which could profit-ably compete. Countries contemplating methanol or fertilizer pro-jects at present need therefore to be cautious and engage in careful economic appraisals. This note of caution does not rule out the possibility of good projects sometimes being found however, and should not stop the search for such opportunities where they do exist.

There is also a need for countries to beware of the rent on their gas resources being transferred outside the country through projects of this nature. For despite, or because of, the depressed world markets for the products, contractors are naturally anxious to create work for themselves. Some, hopefully the exceptions, will make a strong case for projects which are not economic. They sometimes succeed in raising financing packages for say, a fertilizer plant, under conditions where they themselves are ensured of a modest return, while the country takes the risk and gains very little from the scheme. In such cases its gas reserves are depleted to no good cause. Put differently, the opportunity value of gas as a tradable fuel (substitute for, say, imported oil) could well be higher than that of gas inputs in these uses.

In any case the capital cost of methanol/fertilizer plants is high and such investments are risky. Projects of this type enjoy a greater chance of being economic where there is production of large amounts of associated gas with no alternative use, where gas is produced at very low cost, or where there are no capital constraints. The existence of a local market for the products may also help. The oil-exporting coun-tries of the Gulf fulfil *some* of these conditions; but there, high labour costs, organizational constraints and the absence of local markets cast a shadow over the profitability of such projects. Large agricultural countries such as India may provide a local market for fertilizers; but even there the promotion of fertilizers relates to complex agricultural development policies, and there are competing uses for the gas.

(*e*) *Compressed Natural Gas.* Some excitement can be engendered around the potential for using natural gas in 'non-conventional' markets. Compressing gas into cylinders makes it transportable in small quantities, thus opening up new market opportunities. The technology for this process is well developed. In New Zealand, for example, over 100,000 vehicles are fuelled by CNG.

There is indeed scope for use of CNG in the developing world; but only in limited circumstances. In particular, it is unlikely that CNG applications would ever account for very large volumes of gas use. They may provide extra bonuses, where a pipeline already exists, but will not in themselves justify the development of a gas field or the building of a pipeline. In some instances CNG may be usefully distributed to small businesses, such as brickworks or cafés, which do not merit spur lines from the main gas transmission network, but which are easily accessed by vehicle from the main road, which often runs alongside major pipelines. Such usage would always depend on the relative price and availability of LPG and similar substitutes. A much wider application is likely to be in the transport sector. In many developing countries the consumption of gasoline and especially diesel is a major worry to the government because of its impact on the balance of payments. Conversion of vehicles to run on CNG involves considerable capital cost; but the payback period for conversion will of course depend on the price of CNG relative to gasoline and diesel. In New Zealand for example it is estimated that conversion of a petrol-powered passenger car (less than ten years old) which travels an average of 7,000 miles per year has a financial payback period of approximately two years. In addition to the issue of relative prices, it should be recognized that few vehicles in developing countries will meet such criteria of age and annual mileage, apart from some taxi fleets. Furthermore the provision of filling facilities for private vehicles is relatively expensive. The real potential seems to lie with the diesel fleets of buses and trucks, which account for a large proportion of road traffic in many developing countries. In many cases these are concentrated in the urban centres of the country, and tend to make short journeys around a central depot. They are ideal candidates for conversion, since they can refill at a small number of places, often overnight; this allows for the cheapest filling technology to be used. Furthermore, it is often diesel consumption that creates the worst economic problems for a country; for local oil refineries often produce a surplus of fuel oil and are unable to produce sufficient diesel, leading to the need to import large quantities of the most expensive oil fraction. Substitution in this sector could therefore bring real economic gains to a country. It will always be necessary to weigh the relative advantages of using CNG or LPG in vehicles however, depending on availability of the two fuels and the relative costs involved.

(f) LPG and Condensate Extraction. Where natural gas contains a high proportion of condensates or propane and butane the opportunities for developing a gas field can be much greater than when the gas is dry. The chief reason for this is that both are internationally tradable commodities, and can thus assure the oil company of income in hard currency. Condensate extraction (for export) was a major attraction for Unocal in Thailand for example. Meanwhile from the country's point of view the LPG is an extremely useful product. It is much cheaper to bottle and transport than CNG. It can therefore be more easily distributed to rural areas and can meet demand from domestic and small commercial customers where a gas distribution network cannot be justified. The Thai government is using the LPG produced with its gas and condensates to enhance supplies to rural areas.

Most gas fields however do not contain a high enough proportion of these products to allow for their production where there is no market for the dry gas. Taking the example of Thailand again, there is a shortage of LPG; but the government could not build the second gas separation plant it originally planned, until a market was found for the additional methane which would thus become available. LPG and condensates may therefore be a useful by-product of natural gas, but are rarely the engine for field development. At most they may provide the finishing touches to a package which can attract an oil company.

1.5 The Oil Price Collapse and Gas Development

The major competitor for gas in most of its potential markets is oil. Even where coal is also a potential substitute, the economic baseline tends to be set by the price of fuel oil. Which gas fields are developed, in which countries and for which markets, thus depends ultimately on beliefs about the future price of oil. From 1974 to 1984 the high price of oil, and the perception that the price would continue to increase, gave a large economic incentive to the development of gas reserves (whether or not such reserves were actually developed). The collapse of the oil price in 1986 changed the perceived economics of gas schemes however. It became legitimate to ask whether the whole issue of gas in developing countries had not suffered a major set-back.

Low oil prices not only intensify the competition between oil and gas, making expensive gas fields uneconomic to exploit. They also reduce oil company cash flow, restricting companies' ability to invest in new ventures either through the use of internally generated funds or with funds raised from outside sources. At the same time the financial sector in general, and international institutions such as the World Bank which have in the past encouraged gas development, are strongly influenced by the short-term situation of the oil market, and are changing their tune.

Investment decisions should however be made not on the basis of current prices, but on the expected prices for gas and oil when the investment matures. Even when there are no extrinsic delays the development of a gas field in a developing country always implies a wait of three or more years, while infrastructure is built, a market promoted, and so on. Forecasts of oil market movements are therefore crucial to decisions over gas investments.

The conventional view of future oil price developments is in fact relatively optimistic for gas. According to this view, a tightening of the oil market is expected in somewhere between four and ten years' time. Low oil prices will sooner or later stimulate demand, both through a general increase in consumption and through lower incentives for fuel conservation measures. Meanwhile restricted cash flows and hence investment in the oil industry (and other fuels) will limit future production capacity. These two factors will produce a tighter market, leading to increased prices. The world distribution of oil resources will further accentuate this trend. It is likely that oil exporters outside OPEC, such as the UK, Egypt and possibly the USSR, will cease to be net exporters sooner than those within OPEC, probably some time in the 1990s. This structural change on the supply side will coincide with the tighter market described above. The competition of the late 1970s and 1980s, which eroded OPEC's power to hold the price high, will thus reduce significantly, with power concentrated in the hands of the few remaining oil exporters. The tighter market will thus be dominated by a smaller number of oil exporters – basically the Gulf, including Iraq and Iran.

The conventional scenario is thus that the world will face a new era of high oil prices. The outlook for gas in this scenario is good. High oil prices provide ample scope for gas to substitute for oil in most sectors other than transport. A technological breakthrough in the form of a cheap process to transform natural gas directly into gasoline or to diesel would put a heavy lid on oil prices, but would also open up to natural gas the huge premium market of the transport sector.

Alternative scenarios propounded by dissenters from the conventional view are not so happy however.[4] These scenarios postulate low growth in world oil demand over the next ten years, due to restricted economic growth, continued subsidization of non-oil fuels in certain industrialized countries (e.g. coal in West Germany) and further gains in energy efficiency as the benefits of investments and technical advances initiated in recent years are fully realized. These scenarios also postulate a continuing increase in non-OPEC oil supplies, thanks to the cumulative effect of small additions to reserves in a large number of countries, and to further investments in the North Sea, North America and the Soviet Union. Low oil demand growth and increases in non-OPEC supplies would squeeze OPEC production and prevent significant oil price rises. In this scenario the prospect of gas developments in the Third World would be adversely affected by a growing perception of future weakness in oil prices.

In summary, then, there is little doubt that the 1986 oil price collapse has reduced the opportunities for gas development in the Third World. However, the conventional view about future oil price developments is favourable to gas investments in countries with low-cost reserves and reasonable market prospects. Development in the next few years will depend critically on whether the current conventional view on future oil prices retains its credibility, or whether it is displaced by the pessimistic perceptions of alternative scenarios.

[4] The most prominent is Francisco R. Parra, who recently expressed pessimistic views about the future growth of oil demand and about the residual call on OPEC oil at the ninth Oxford Energy Seminar.

2 GOVERNMENTS AND COMPANIES

2.1 Introduction

This chapter looks at the two agents who are involved in a gas project, namely the company and the government. It sets the framework for later discussions on details of the negotiation of contracts, by analysing the frame of mind in which each party approaches such negotiations. For both company and government we ask three questions:

(a) How interested are they in gas in principle?
(b) What are their objectives – what do they hope to get out of a gas project?
(c) How constrained are they by shortages of finance and skilled manpower?

We then look briefly at how the divergence between their respective aims and objectives affects both their relationship with each other, and the conduct of negotiations. Still looking at the relationship from a broad perspective we also examine the ways in which negotiations can be complicated by the effects of corruption.

2.2 Companies

(a) *Their Interest in Gas.* This section will discuss the nature of companies' interest in gas, and the problems they perceive in developing their gas reserves. Existing and future gas schemes in developing countries all involve oil companies.[1] In some cases the national oil company has exclusive rights over development, but often foreign companies and multinationals are involved. A very large number of companies have in fact discovered gas reserves in the South. In the eight countries covered by our case-studies the companies that have found gas include the following: AGIP, Amoco, BP, Burmah, Elf, Exxon, Shell, Texas Pacific, Total, Union Texas and Unocal. While

---

[1] The gas utility companies of Europe and the USA have not so far been directly involved in upstream gas activities in developing countries. Their interest has been limited to the occasional role of consultant on downstream projects; for example British Gas planned the residential distribution network in Cairo.

most of these companies do have substantial upstream gas interests in both the industrialized and the developing world, they are nevertheless seen principally as 'oil' companies.

There are certain attitudes about gas development which oil companies in general are widely believed to hold. A frequently stated assumption is that oil companies are not really interested in gas: that they rarely look for gas for its own sake, and only come across it as a by-product of oil exploration, when it is viewed as a positive nuisance. This is certainly not universally true. First, there are differences in companies' attitudes as regards gas development in industrialized and developing countries. Secondly, there are differences in approach between companies.

Oil companies react very differently towards gas in the Third World, compared to their response in industrialized countries. Indeed, they have actively explored for gas in developed countries, and many have been instrumental in promoting its development through subsidiaries or through the utilities. It is undeniably true however that they find gas in developing countries to be problematic. One of their chief worries is that should they find gas in developing countries then the government will sooner or later put pressure on them to develop the resource, regardless of their own judgement about its economic viability. In recent years additional pressure has come from international aid organizations to pursue projects which are economic, but which may still seem relatively unattractive to the company. Some companies have responded by avoiding the occasion of such pressure, namely reducing exploration in gas-prone areas.

The factors that disincline them to develop such gas reserves are developed at length elsewhere in this book. But briefly, three of the most commonly perceived problems are:

(a) The lack of a contractual/legal/fiscal framework specifically designed for gas.
(b) Even more importantly, the lack of potential markets for the gas, or the need to promote markets where none currently exist.
(c) The lack of flexibility afforded them when developing a gas reserve for a local market (as opposed to international trade in oil), and the problem of foreign exchange convertibility, that is, of repatriation of some of their profits.

Although there is much in common in the behaviour of oil companies, there are also important differences between them. Some of the most important differences arise from the following characteristics:

(a) the size of the company;
(b) its historical exposure to overseas involvement;
(c) its company/management philosophy;
(d) the degree of experience and skills in gas that it already has;
(e) its nationality.

Some of these are obvious in their implications. The tendency of one company to spread its interests widely, to build up its ownership of hydrocarbon reserves and contacts for the future, while another may seek more immediate returns on its investment, can have a great impact on their willingness to explore in developing countries. Furthermore the influence of the company's nationality as such is often noticeable. For historical and political reasons particular companies have very strong links with certain countries, such as the French oil companies in French-speaking Africa, and may be less disinclined to invest there than other companies. The US government gives very strong, but not always conspicuous, backing to its companies in certain countries. This probably influences the behaviour of US companies as compared to say European firms which receive strong political backing from their own governments in other parts of the world.

Some companies recognize more explicitly than others that gas exploitation in the Third World is an important undeveloped area of business, which could provide an important growth area for the future, providing the problems mentioned earlier can be overcome. This is particularly marked in some of the largest oil companies, which are able to take a long-term strategic view of the world. Long-term planners in such companies are increasingly beginning to worry about the implications of the imbalance in their hydrocarbon reserves. Following the run of nationalizations by OPEC countries in the 1960s and 1970s most of the majors today own only a small percentage of world oil reserves, compared to their share in the world market for petroleum products. Furthermore their reserves/production ratios have been declining for some time. This is an inducement to them to search for new reserves outside the two conventional exploration areas of OPEC (now largely closed to them) and North America and the North Sea (where exploration efforts are already beginning to show signs of diminishing returns).

Some companies, particularly the majors, are therefore beginning to think in terms of positively looking for gas as such, since they are concerned about increasing their total reserves, and the largest increments at the margin are likely to be gas.[2] At the same time this leads

[2] Indeed, a feature of recent exploration results has been that more gas than oil has been discovered, even when looking specifically for oil.

to a preoccupation with what to do with the gas they may find. The impact of this type of strategic thinking should not be over-exaggerated however. Certainly it is still confined to three or four of the very large companies. Most other companies are immersed in today's other pressing problems, caused by the collapse of the oil price, and are restricting their outlook to a very short time-horizon. But once these pressures are eased the indications are that the attention of more companies will move in the directions indicated above.

(*b*) *Company Objectives*. It is important to emphasize that in the first and last analysis an oil company is simply a private commercial concern, whose dominant consideration is to generate profits. Of course the modalities of pursuing the profit motive can vary, and other objectives can also enter the picture. But in most cases such additional objectives will be either directly or indirectly related to the profit motive. For example the company may emphasize its wish to grow, or to increase its market share. But growth cannot be sustained without profit; and pursuit of a larger market share, although it may occasionally prevent an immediate increase in profits through loss-leaders and so on, is merely a way of ensuring long-run profits by defending the market for its products. Profit-making always remains central to a company's behaviour.

Profit-making is however an inherently risky business. A company's success depends on striking the right balance between potential re-wards and the risk involved in achieving them. Large profits can result from taking a calculated risk; but at the same time the company must not over-expose itself. This very general stricture applies both to the corporation as a whole, and to decisions over particular invest-ments. As discussed in 2.2 (*c*) below many companies are currently much more risk averse than they have tended to be in the past, and this is colouring all their investment decisions. But in this section we will concentrate on what specific decision-taking mechanisms are used by companies to balance the upside potential of a project against the risks it carries with it.

On a company-wide scale, one method of reducing risk is to diver-sify the company's business. In this context gas in developing coun-tries offers some useful openings to oil companies. Diversification can take a number of forms. It may be spatial, across different regions and countries – for many oil companies involvement in the South would provide a balance to over-concentration in the North Sea and North America, as mentioned earlier. Diversification may also be across different activities: during the 1970s a number of oil companies moved into fields as diverse as nuclear energy, business machines and mail-order firms. Many of these schemes turned sour, and we are now witnessing a retrenchment of oil companies away from such wide-spread interests. Diversification into gas in developing countries on the other hand would offer an opportunity closer to the companies' traditional skills and knowledge, and hence should be of interest to them. Diversification may also be over time. Some companies show interest in this, working to build up long-term potential, whereas others concentrate more on short-run returns. For example Shell is gradually building up footholds in a number of new countries, some-times with more gas than oil, with a view to long-term returns.

When considering investment in a specific project however, a com-pany will expect it to meet in its own right company criteria of profitability and riskiness. Only if it can meet the required standards will it be of any use in diversifying the company's overall business.[3] The decision to go ahead with a project will be arrived at through a combination of predictive techniques and evaluative judgements. The factors which determine profitability can be measured: the return on investment can be calculated through a discounted cash flow exercise; the pattern of cash flow over time can be drawn up; and the payback period assessed. But ultimately the results of all of these apparently precise techniques depend on uncertain future events, hence on the company's assessment of the risks involved in a project. It should also be borne in mind that a project with great upside potential will be inherently more attractive, even if the risk is relatively large, than one which will allow only modest but safe returns.

When considering investment in gas in developing countries com-panies encounter specific risks. We next consider what these are, and what routes, if any, are open to the companies to mitigate their effects. Since the payback period for gas investments is typically quite long – often longer than for the companies' accustomed oil business – there is a tendency for arguments to always revert to the expected rate of return on gas projects. A higher cut-off point for acceptable rate of return or a higher discount rate are often used in assessing gas as opposed to oil projects.

Exploration Risk. Oil and gas exploration is an inherently risky under-taking, with companies inevitably investing in many areas which prove fruitless. The industry as a whole therefore seeks a higher rate of

---

[3] In some situations this may be qualified; a loss-leader may be acceptable for instance if it is a necessary condition for other, greater, profits to be made. It is not inconceivable that an oil company would be prepared to invest in gas in order to gain the goodwill of a country, if this would enhance the company's prospects of present or future activities in upstream or down-stream oil.

return than is common in other industries, and the industry's usual 'defence' for this approach is that each successful well has to cover the costs of many dry holes.[4] Furthermore the companies tend to look at this feature on a world-wide scale: the fact that some countries tend to have better-than-average success rates will be used to balance the poor record of other countries, not to lower the required rate of return within the successful countries. This may of course be unpalatable to governments, who feel that they are subsidizing the companies' activities in far-off lands. But there is little they can do to change this state of affairs.

'Political' Risk. Companies tend to be wary of all investments in developing countries. Contrary to first appearances this is not only because of 'political' risk in a narrow sense, that is, a fear of political instability and changes in regime. What companies claim is of great concern to them is the broader problem of a given government changing its policies, especially its fiscal and taxation policies, either erratically, or consistently in an unfavourable direction.

Companies tend to react strongly against abrupt and unpredictable changes in government policies, even when their business interests are not in fact harmed by the changes. On the other hand, and paradoxically, if policies are relatively stable companies adapt quite readily to meet changes in circumstances which make their business rather marginal. This is especially true once a company has become entrenched within a country, and has built up its network of contacts. The case of Burmah Oil in Pakistan is a case in point; the company continued to produce gas for many years despite the fact that its rate of return came to seem very low compared to that required for new projects.

Unwelcome policy changes are alleged to be more of a problem in the Third World than elsewhere; but the facts do not bear this out. The UK government has a history of making very frequent changes in taxation provisions for North Sea oil. In contrast the Egyptian government has been scrupulous in respecting both the principles and modalities of its oil production-sharing regime over a very long period, including a time when the companies were making unforeseen profits. In this respect investment in the North Sea has been much riskier than investment in Egypt.

---

[4] Professor Edith Penrose in a comment to the authors stated that she has never been convinced by the argument that oil and gas, because of exploration risks, require a higher rate of return than is the case for other research-based international industries, such as the ethical drug branch of the pharmaceutical industries. We wholeheartedly agree. It remains true however that research-based industries will usually seek higher returns than non-research-based ones.

Indeed the kernel of the problem seems to lie not in policy changes as such, but rather in the company's perception of its influence over the government. In the UK companies are familiar with the lobbying system, and even if unwelcome changes are made in the fiscal system (such as the PRT proposals of November 1986) the companies feel there is room for them to make their views heard, and for the system to be modified in their favour. In developing countries, on the other hand, the companies feel themselves to be outsiders. Even when they employ local lawyers they can never attain the same influence over officials, or rely on such effective (often informal) communication channels. They do not know how to 'work the system'. If unfavourable measures are introduced they may feel they have no other option than to leave the country; whereas in Western countries they would be confident of their ability to overcome short-term differences of opinion with the government.[5]

Market Risk. Companies are concerned that investments should not be too rigidly dependent on particular customers behaving well. Unfortunately the character of gas projects in developing countries requires large, lumpy investments in schemes where there is a rigid physical link between the gas source and the market, and where the company is tied into rigid bilateral relations with one customer, the government. From the point of view of alleviating market risk the situation could hardly be worse. The company will need to become closely involved with the government over a very broad range of issues. No other investment by a multinational requires negotiations and agreement with the government over such a broad front. Moreover the initial terms of the understanding will need to be applied, and adapted in acceptable ways, over a period of about twenty years, if acceptable returns are to be earned on their investment. This is a unique situation for the company to find itself in, and is naturally perturbing. The company finds itself heavily dependent on the government to maintain both prices and offtake. Initial prices may be acceptable, but may involve a financial subsidy, where consumer gas prices are much lower than the purchase price from the companies. Such an arrangement makes the companies nervous that a future downturn in the government's fortunes may leave it unable or unwilling to continue the subsidy. Similarly, take-or-pay clauses may be agreed, but it is commonly accepted that such clauses will not hold up if there are important changes in the economic environment such as

---

[5] Caltex is an interesting counter-example of a company with a strong Third World orientation. However Caltex is owned by two majors which have considerable investments and activities in the industrialized countries.

the recent oil price collapse; they function more as a signal of initial goodwill than as a long-term guarantee.

Since the number of outlets for gas in a developing country is normally very small, the theoretical possibility of a 'free market' is in reality very limited. Nevertheless the companies will always try to encourage moves in this direction – not so much as part of an ideological crusade, as from a purely practical desire to insure their profits against market risk. Occasionally gas does offer other marketing opportunities, and these should be taken full advantage of. A case in point is if the gas happens to be rich: for example in Thailand Unocal has been able to export its condensate, or sell it close to international prices to the government. Such apparently insignificant features can make or break a project. If there is genuinely no scope for increasing flexibility, then the company will be anxious that the project is not unduly risky in other respects.

Convertibility Risk. To earn a profit is not enough in itself. If the investor is a foreign company, it will need to be able to repatriate some proportion of the profits. If the scope of its activities in a country is large enough a firm may be able to plough back most of its profits into the economy where they are generated for a limited period; but ultimately it will need to distribute those profits to its shareholders. For oil export projects there is of course no problem, since the products will earn hard currency. The same applies to LNG, and to methanol or fertilizers (for export), which may be produced from natural gas. But for internal gas (or oil) utilization, where only local currency will be earned from sales, a considerable difficulty emerges. The problem is of course well known. Very many developing countries are suffering from a severe debt crisis, which leads them to impose very tight exchange controls. A few countries are exempted from this problem, such as Thailand. But some of the largest countries, which would otherwise seem most attractive for gas developments because of the size of their potential gas markets, are suffering most severely – for example, Argentina, Brazil and Egypt. Smaller countries which do not have such large debts still operate under very strict exchange controls. There are ways of assuring the companies of profit repatriation: third-party guarantees, the government setting aside 'import savings' in an escrow account, and so on. The idea of an escrow account seems rather unrealistic however; existing creditors would demand first claim on such funds, and the government would be forced to give low priority to the shadowy notion of 'import savings'.

Alternatively it has been suggested (and practised in Egypt) that companies could be paid in exportable crude oil for the gas they sell within the country. But again there is a problem. The policy would be difficult to carry out in the case of a company which happened not to be already producing oil in the country. In any case crude oil is in fact equivalent to foreign exchange in such a situation – by giving away more oil the government reduces export earnings. The advantage of such a system is that the national oil company has it within its own powers to release oil as payment, whereas it often needs to go to the Central Bank or Ministry of Finance to obtain foreign exchange. It is therefore administratively easier and often much speedier.

Despite the problems in overcoming convertibility difficulties it is vital that its full importance is recognized by the countries. For without convertibility of earnings the very foundation of the companies' initial interest in investment, the opportunity to make a profit for its shareholders, is vitiated. The only path which may carry some hope in this respect is the extension of schemes offering third-party guarantees, such as MIGA. This is discussed in detail in Chapter 4.

(c) Financial and Manpower Constraints on Companies. Manpower constraints are not normally seen as a major limiting factor on the companies, but may in fact play an important role. It is true that most oil companies have experience with gas in the developed world (though mainly upstream) and have staff with the requisite skills and experience. But there is a limit on their absolute number, and hence on the total number of projects the company can undertake. Companies will also be anxious to avoid unproductive use of staff, for example in very prolonged negotiations, when they could be more profitably deployed elsewhere in the company's business. This will especially apply to their most skilled negotiators, who are always in short supply. Consequently relatively junior and inexperienced staff may be allocated to the protracted negotiations for a gas project in a country distant to company headquarters. While top negotiators may be brought in at the crucial stages this may still be insufficient, given the need for a very high level of sensitivity to government behaviour and considerable knowledge of political economy. Manpower may therefore be a serious limiting factor during the crucial early stages of a gas project.

Finance presents another serious problem. In the past it was widely assumed, and correctly, that if an individual project was ranked high enough in an oil company's investment priorities then the company would pursue it. Today this is no longer true. Company negotiators are heavily constrained by over-arching corporate considerations quite outside the actual project under scrutiny. The US oil industry in particular is going through one of its worst periods this century.

The reasons for this are a combination of falling oil prices and US corporate trends. The fall in the oil price, together with its uncertain long-term outlook, have had a tremendous impact both on company profits and attitudes. Corporate cultures which had lasted for decades have been jettisoned. Caution, improved efficiency and cost reduction have become the new management bywords. This effect has been magnified by the concurrent spate of take-overs which has character-ized the industry in America. During the 1980s a number of com-panies (such as Unocal and Phillips) have taken out huge loans in order to fend off take-over bids; and this has led to a sharp increase in debt/equity ratios. Recent reductions in the valuation of oil and gas assets have further added to the effect. The immediate result of this has been to reduce cash flows significantly, and to restrict options and manoeuverability. Many companies are now preoccupied with improv-ing short-term results and cash flows, in order to protect their finan-cial ratings, or in the case of those companies with high debt/equity ratios to lay the groundwork for improvement.

In the context of gas projects this trend has two important implica-tions.

First, companies have made drastic reductions in expenditure on exploration and development. The *Oil & Gas Journal* (March 23, 1987) reports a 31.4 per cent average decrease in capital and explora-tion expenditure by twelve large companies in 1986, as compared to 1985. This ranged from a 13.6 per cent decrease for Mobil to as much as 69 per cent for Amerada Hess. Other major companies in the group reporting decreases included Shell (36.2 per cent), Exxon (33.1 per cent) and Amoco (41.4 per cent). Further reductions are expected in 1987, with Exxon planning an additional 10 per cent reduction, and Arco as much as 20 per cent. The net result of this pattern is that countries are finding they need to compete hard if they wish to attract scarce funds. In an effort to move up the companies' priority listings they are offering larger shares of project earnings and trying to reduce risk to the companies.

The second result of the new corporate climate is that companies have become extremely wary of marginal projects, especially if they involve heavy front-end investment as gas normally does. Future business activities will give priority to projects able to pay their own way in a very short period of time. Rate of return has become far more important than increasing market share. As a consequence, com-panies are becoming much more averse to taking on risk, and will be much tougher in negotiations for new ventures. Their decision criteria will be even more stringent than they have been in the past.

(*d*) *Company Summary.* All companies are motivated primarily by the need to make profits for their shareholders. They will vary in their corporate strategies to achieve this end, since their perceptions of risk vary, and their hydrocarbon reserve portfolios also vary. But in principle the generalization can be made that if companies see a potentially profitable opportunity to develop gas they will be intent on negotiating an agreement with the government which will minimize their risks and maximize their potential profits.

In the current commercial climate they are also particularly keen to avoid very risky ventures. Although gas projects in developing countries can improve a company's risk profile, in the sense that they diversify its interests, they also carry with them a range of specific risks which companies are reluctant to take on. The most important of these are the fact that gas projects require large front-end investments which depend for their profitability on the long-term goodwill of governments over which they have little influence; and companies are likely to experience problems in repatriating their profits. A large part of this risk is thus a result of the lack of trust between company and government.

We now turn to look at the government's view of the problem.

2.3 Governments

(*a*) *Their Interest in Gas.* It is assumed that in principle governments are interested in developing the natural resources of their countries. Furthermore, apart from exceptional cases, governments throughout the world tend not to worry too much about the depletion problems of an exhaustible resource (such as oil or gas), particularly when the resource has already been developed. They have a short time-horizon, with a high discount rate for the future. This is accentuated in the case of developing countries with their pressing immediate needs and limited resources.

These observations and assumptions should not lead us to conclude too quickly that a government which knows that the country has, or is likely to have, gas reserves will necessarily take immediate and effective action to promote the development of the resource. There are differences between countries and changes in the behaviour of individual countries over time.

Historically, there are instances of observed apathy. Governments of oil-producing countries witnessed the flaring of their associated gas over many decades. Sometimes they complained vehemently about 'waste', but often they accepted the companies' view that associated gas was a by-product with no economic value. Even in recent times when the rise in oil prices clearly indicated the high opportunity value of gas, they were often still apathetic. In some cases this was due to ignorance about gas matters, or to gas being treated as the neglected 'little brother' of beloved oil which attracted all the attention. In others, it was due to a realization that, as far as gas was concerned, the country was 'demand constrained'. There are also perverse instances where vested interests in some part of the state system (often a national oil company) blocked the progress of gas out of fear of competition with oil (e.g. Brazil).

In contrast, there are also instances of active interest in gas development. In some countries official interest in gas has a long history (Argentina, Pakistan), while in others it was aroused by the rapid oil price rises of the 1970s, or by the prospect of declining oil reserves and decreased export revenues (Egypt, Tunisia, Malaysia, Indonesia). Attitudes towards gas have also changed: the flaring of associated gas that had in some cases continued for decades has in many instances been greatly reduced.

There are good reasons why governments sometimes hesitate over gas development. First, the potential benefit of gas development to a country arises from turning, at a cost, a natural resource left idle in the ground into a fuel which if used domestically can either release oil for export or substitute for oil and other fuels in the import bill. The potential benefit of gas development depends therefore not only on the costs involved but also on the future prices of substitute energy, which must always be uncertain. For a country, the decision about gas development (as for other exhaustible resources) involves a choice between two risky options – inaction and action. Assuming that costs of development are correctly assessed, the risk in both options is due to the same cause, uncertainty over future prices. If we recognize that action involves additional risks – costs may run ahead of initial estimates, demand and actual market size may fall short of expectations, etc. – one can understand why the scales may sometimes tip in favour of inaction.

But even when gas development appears to make economic sense (expected benefits higher than expected costs, properly measured and discounted), governments often appear to hesitate and delay. This is partly due to the constraints associated with underdevelopment as such, namely lack of skilled manpower and of finance. It is also the result of the very complex web of aims and objectives which governments have to try to satisfy at any one time.

Secondly, governments often hesitate because the macroeconomic implications of gas development are considerable. This is basically due to the fact that it typically involves very large upfront capital expenditure, not only upstream but also in the transmission and distribution system, not forgetting also the costs to consumers of converting from one fuel to another. For a typical Third World country a large investment in the gas industry competes for scarce domestic resources and outside funds, not only with alternative energy projects but also with all other worthwhile investments in the country. The macroeconomic implications are not limited to issues of raising finance and controlling national debt. Through the pricing issue gas utilization impinges on the whole problem of inflation. Furthermore, gas development may also in certain situations raise socio-economic and regional problems.

Finally, there are also factors inherent to governments as institutions which can cause hesitation over promoting gas development and delays when considering projects. In particular governments always have to satisfy a much more complex web of aims and objectives, as compared to the relatively straightforward objectives of companies. They also suffer considerable constraints on skilled manpower and on finance. The remainder of this section will explore these aspects of governments, and their impact on gas projects. Section 2.4 will then bring together our analyses of companies and of governments and look at the interaction between these very different bodies when they attempt to negotiate with each other.

(*b*) *Institutional Structures and Decision-making*. Before moving on to discuss the objectives of government we need first to look in more detail at what is meant by 'the government' in the context of this book.

The government/public sector set-up with which companies have to deal for gas development is not a homogeneous entity, due to the number and complexity of issues involved. The structure of responsibility for gas issues, and the division of this responsibility between different government institutions differs from country to country. In most countries a number of ministries will be involved. Apart from the ministry in charge of oil and gas, the ministries of planning (or economic affairs), industry, and electricity may all play a role. Usually in the background, but with an important role, there will also be the Ministry of Finance and a Central Bank. In addition, in many of the countries covered in this study there is also a national oil company which may be involved in exploration (EGPC in Egypt, PTT in Thailand, YPF in Argentina, etc.) and sometimes a subsidiary or separate company responsible for gas sales (Petrogas in Egypt, Gas del Estado in Argentina, etc.).

That the responsibility for energy policy is distributed among a number of institutions is not pure caprice however. Nor is it peculiar to the developing world. It results partly from history, partly from the complexity of the issues involved, and partly from the fact that the state is simultaneously producer, financier, manager, regulator and policy-maker in the wide socio-economic areas affected by gas development.

The fact that the government/public sector is not a monolith but rather a set of institutions means that each may have its own perceptions and objectives, and there may be a conflict of interests to be resolved.[6]

Conflicts may arise between any of the agencies involved. A typical case is between the oil/gas authority (whether this be the ministry or national company) and the agency in charge of electricity. This could either be over the scale of conversion to gas, or over pricing and take-or-pay arrangements (if the latter are contemplated). In Thailand, for instance, EGAT was keen to revert to burning fuel oil in the power stations when a large price differential appeared between fuel oil and gas; but PTT was committed to a take-or-pay arrangement, and prevailed on them to continue to use gas. Other ministries may also become involved however. In Indonesia, a division within the government appeared over pricing. A government decree set gas prices at the plant gate, but then the Ministry of Industry began to make exceptions. For example Mobil was told to sell gas at one-third of the gate price to a pulp and paper company. Pricing thus became subject to political infighting between the Ministry of Mines, which wanted higher, and apparently unreasonable, prices, and the Ministry of Industry which wanted lower prices. Again in Thailand, problems have emerged between PTT and the Planning Commission, not only because the latter is vested with authority to approve/reject all public investment proposals, but also because the planning agency is promoting a privatization and anti-monopoly philosophy which threatens PTT's role and existence.

A further instance of possible conflict is between the ministry for energy and the national oil company. In some countries such differences are internalized because the company is virtually integrated

---

[6] Of course there is also a plurality of interests and perceptions within large companies, whether the corporate structure is centralized, as for example in Exxon, or structured in the form of semi-autonomous profit-centres, as in BP. But most companies do not allow such differences to appear too blatantly in their dealings with outsiders (except perhaps in trading operations on the world market). When decisions have to be made the conflicts are resolved by higher authority. Furthermore, the problem is simpler for a corporation than for a government because the objectives of the former are straightforward, and well understood by all those involved, while the political objectives of governments are complex, and are subject to different and ever-changing weights.

into the ministry, with little autonomy. In Iraq, for example, it is difficult to distinguish between the INOC and the Ministry of Oil. In Egypt the Minister of Petroleum was until mid-1987 also the Chairman of EGPC. Under such circumstances conflict becomes wholly internal, and a unified face is presented to the outside world. Where the national oil company is autonomous from policy-making bodies the differences in approach become more public. In Argentina, for example, the Secretariat for Energy finds the task of supervising YPF particularly frustrating because of the very entrenched position of the state corporation. Paradoxically, international oil companies prefer national oil companies to be clearly separated from the policy-making body for operational reasons. But if the two 'government' bodies cannot agree between themselves on a negotiating stance this obviously causes problems for the companies.

An additional element of institutional rivalry can enter with the advent of gas. Interest in developing gas is quite a recent phenomenon, and new and inevitably junior institutions have sometimes been formed to carry out the task. These have often split off from a pre-existing oil enterprise. While this is a reasonable procedure, in the sense that gas requires different treatment to oil, particularly in relation to the need to promote a market, it does tend to lead to tensions. This is exemplified in the case of YPF and Gas del Estado in Argentina, where YPF can see that the future lies with gas, and is therefore on the defensive, while Gas del Estado in its turn is afraid of being reabsorbed into YPF.

Why is there such a divergence of interests? The causes are many. First, there is competition over financial resources. The general fiscal crisis of the state in most developing nations means that there is a great effort to force national enterprises (be it for oil, gas or electricity) to cut expenditure; while the ability of the state enterprise to raise revenues is nearly always conditioned by the government's pricing policies. Typical examples are PEMEX in Mexico, and PTT in Thailand. A similar pressure can be seen in Malaysia, where Petronas is determined to push for the construction of a gas pipeline, against the government's expenditure policy. Secondly, the national enterprise, as any institution, seeks to establish its own identity and in the process develops its own interests, which can be different from those of the supervisory ministry or the government. Careers, prestige, the motivations of managers and their 'satisficing objectives' (which may relate either to personal income or the growth of the firm) are all involved. Thirdly, there are often group, if not personal, rivalries between civil servants and officials of national enterprises. The former have supervisory power but the latter manage real resources. Access to resources elicits envy, and the exercise of power can cause frustrations. Fourthly, and most importantly, the conflicts which may arise between national enterprises and ministries (or other organs of government) are due to inherent differences in their respective functions. The national enterprise is a specialized agency with well-defined attributes and responsibilities. It has specific and concrete tasks to perform in a narrow field of activity. The political agencies of governments, by definition, have a different and wider remit. They are primarily concerned with policy on a very broad front and with the trade-off between numerous objectives. In these respects, the national oil company is closer to international oil companies than other government agencies are likely to be. It shares the same operating philosophy, engages in the pursuit of profits, and wants the job to be done.

Such conflicts, when they exist, naturally affect the smoothness of decision-taking, and cause delays. Consider the common instance in which a multinational is negotiating with a state enterprise. The foreign company will soon discover (if it did not already know) that many elements of energy policy and other important decisions lie outside the remit of the national enterprise. It is frustrating to negotiate with an agency which is not able to settle many of the items on the agenda; and the frustration is further compounded when it is impossible to identify which authority does have the final say on various matters. (There is rarely a single authority with decisive influence on every matter.) In this context, the balance of power between different ministries, committees and official agencies, the relationship between the state enterprise and the other state institutions, and its ability (as the primary negotiating body) to communicate the foreign company's views to the relevant ministries and agencies, become crucial.

The difficulties discussed in this section are essentially a 'state political' problem; they arise from the very nature of the state sector, and are not peculiar to developing countries. They must be distinguished from difficulties due to poor organization, bureaucratic inertia or inefficiency, and lack of experience or competence, which may affect countries in different ways and to various degrees. The remedies for these two distinct sets of problems are different. The latter type of problem can often be remedied in time, through organizational reforms, designed with advice and help from management consultants or public administration experts, or through the eventual accumulation of knowledge and experience. The former problems, those arising from the nature of the state sector, call for more original remedies. Our proposed remedies fall into two parts.

First, to set up the following institutional structure which we believe is favourable to gas development. The exploration phase should fall under the remit of the national oil company, given the close relationship between oil and gas exploration, and the fact that a shared philosophy and outlook with international oil companies will ease negotiations. At the other end of the chain should lie a distribution company, whose tasks are most similar to those of an electricity supply company, or other public utilities. Between the production and consumption of gas two distinct functions are required. First, there is a need for a transmission company, which may be either publicly or privately owned. This company acts as a bridge between the upstream and individual customers (whether these are large consumers, such as a power station or fertilizer factory, or a distribution company which in turn serves a large number of small consumers). The transmission company owns and maintains the pipelines; it acts as intermediary should there be several suppliers of gas; and it also carries an important risk-sharing function – it spreads marketing risk, since selling gas to the end-users becomes its responsibility, and it carries the risk associated with loading factors. Finally, there is a need for a regulatory body, which is removed from the immediate need to earn profits, and is concerned with wider policy issues; such a body will normally be part of the government administration.

Secondly, the government may be well advised to appoint a national agency, or even an individual, at the early stages of a gas project, to follow up the progress of negotiations with and between the various official agencies, and to smooth their passage. We call this intermediary the Gas Mediator. Such a Mediator would have the specific task of liaising between the various government interests, ensuring that each institution was aware of its responsibility in the project, and of policy changes that might be required of it. The Gas Mediator would certainly need to have considerable experience of the political scene, and thus be familiar with the personalities involved, the local power structures and rivalries, and ways of doing business. Special skills in negotiation and reconciliation would also be called for, as well as knowledge of the gas industry and enough seniority to command the respect of ministers. Such people are not easily found – one suggestion is that a retired minister might be able to fulfil the role. A Gas Mediator of this kind could help considerably the government's cause. It is however important to stress that such an intermediary would not be intended to usurp the power of other bodies, or to become a permanent institution, as this would simply create a further competing power base, and multiply the bureaucracy even more. The Mediator's role is not to take the decisions which lie within the competence of ministries or the national oil companies, but rather to ensure that such decisions are taken in a timely and co-ordinated manner.

(*c*) *Government Objectives*. Even if the multiple interest groups within a country can be brought together, there still remains a wide difference between 'the government' and foreign firms it may negotiate with. A company is generally interested in economically sound and commercially profitable projects. Governments, in a very fundamental sense, are different entities from companies. They have a wide range of objectives and concerns. They are essentially political entities; and it must be recalled, that for all governments economic factors, economic rationality and economic objectives are often subordinated to the 'higher' realm of political considerations.

This does not mean that governments will not consider the economic merits of proposed developments, but that they will allow other concerns and objectives to weigh in their decisions. And even when the economic returns of a project are the only issue at stake, the appraisal of an undertaking by a public sector entity involves different cost-benefit concepts to appraisals made by firms for the purpose of commercial returns. Governments are expected to be concerned with 'social' as against 'private' or 'commercial' costs and benefits.[7] The difference between these concepts is that the former takes into account externalities (that is, costs and benefits to the society that are not incurred by those who inflict them or not paid for by those who receive them) while these are outside the scope of firms. Typical examples of externalities are environmental damage (for example, pollution caused by an industrial plant, or environmental disturbance arising from the construction of a pipeline) or waste of a non-renewable resource (for example, the flaring of associated gas). Unless compelled by contracts or by law to repair environmental damage or to find a use for by-products such as associated gas a company can ignore these 'external' costs in its appraisal of the commercial returns of a project, although due to political pressure it has often become expedient for them to take into account such externalities in many developed countries. Governments, of course, may also ignore externalities and sometimes do so. But that amounts to a transgression, because they have the duty or the responsibility to take them into account; whereas for a company it is merely an option that may or may not be in its best interest to consider.

Apart from externalities, governments have a number of politico-economic concerns which have an impact on gas negotiations. The

[7] In principle this also applies to national oil companies, although in practice they may tend to act more like private companies.

most important are the following: the country's overall economic and industrial development plans, and the balance of such development between different regions; the distribution of income between groups in the population; the preservation of national independence against possible interference from foreign interests; and vigilance that agreements with foreign investors should not afford the latter unfair benefits. At first glance none of these factors is intrinsic to the development of a gas resource; but in practice they play an important role in defining the government's stance towards companies. We will briefly outline the rationale for these concerns.

Economic and Energy Planning. Economic and energy planning are essential because gas is so bound up in competition with other fuels, and because its development has such wide macroeconomic effects. Both forms of planning have rather fallen out of favour, but this is to the detriment of progress in energy in general and gas in particular. Some countries see planning as superfluous in the light of the succession of shocks that have marked the energy scene in recent years; but the function of an energy plan should not be to provide an eternal and inflexible set of priorities. Rather it should provide the means for coping with uncertainties. The aim of the plan should be to enable the country to meet new and unknown circumstances without causing undue disruption. Several oil companies have remarked to us that life would be much easier for them if the countries only had clear, publicly-stated energy policies. Economic development planning has also become unfashionable. It has become associated with concepts of rigid economic management, inefficient import substitution strategies and an anti-trade bias. But it is possible to have plans which are free from such drawbacks. The need for them is clear in the context of a potentially huge share of national funds being invested in gas infrastructure.

Identification of a market for natural gas also requires strategic thinking about the country's industrialization policy. There is no point in substituting gas for other fuels in industries which are intrinsically uneconomic; still less should ill-advised new schemes be promoted simply because gas is available as a fuel. If the gas is consumed in uneconomic industries the government will be severely constrained in the price it can offer to pay for the gas. Even should the government choose to subsidize key industries for political reasons, companies will be sceptical about supplying the gas, since they will fear future defaults in payment. A clear and concise utilization plan, based on sound economic principles, is of most benefit to the country, and will also encourage the interest of oil companies and financing agencies.

Regional Planning. Governments have objectives of political balance between different socio-political groupings or economic vested interests, between different regions or provinces, between urban and rural areas, etc. Here again political factors are likely to interfere to some extent with the pure economic or commercial appraisal of a gas project. For example, the Nigerian government has put much weight on balancing the interests of different provinces, following the civil war and the vital need to avoid the building up of tension which could once again threaten internal stability. The need to appear evenhanded can then override strict economic considerations. In Argentina other political factors related to regional development have meant that the vast gas reserves discovered in Loma de la Lata in 1978 have still not been developed to their full potential. Here the choice of gas pipeline routes has proved politically difficult, since it will have a major impact on the industrialization of the regions concerned. Consequently the whole project has been repeatedly delayed.

Income Distribution. Most governments, particularly in the Third World, are concerned about income distribution within their countries. They tend however to have a particular bias in favour of certain urban classes which provide political support to the regime or which perhaps represent a potential threat to its security. Political survival is seen as a higher priority than returns to the economy. (Once more, we must emphasize that these considerations do not apply exclusively to developing countries, as anyone familiar with 'election budgets' in Western democracies will readily appreciate.)

In many developing countries the income distribution objective is pursued through pricing policies. In the case of energy, for example, prices of certain fuels are kept artificially low. As discussed in Chapter 3, income distribution would be better pursued through *direct* taxes and subsidies (i.e. through hand-outs to the poor) than through interference with the price mechanism by means of *indirect* taxes and subsidies, because interference with the pricing system leads to an inefficient allocation of resources. Pricing policies cause considerable problems for gas development, and can be a major impediment to the negotiation of a mutually satisfactory agreement between governments and companies.

National Independence. Governments are anxious to protect their political independence, and this may have implications for the relationship of developing country governments with foreign investors. In recent years however they have generally become more relaxed in their attitudes towards foreign investors than in the past. This is a sign both of their increased confidence in their own ability to regulate the operations of foreign investors through appropriate legislation, and an increased recognition of the need for foreign investors to supplement the resources of the developing country in capital, technology and managerial skills, as well as their need to gain access to export markets. Some argue that this increased confidence of governments, particularly in the field of oil and gas, is partly due to OPEC's success in the early 1970s in breaking the hold of major oil companies on the oil economy of OPEC countries. As the oil companies' reactions to events of the 1970s appeared quite restrained, governments which previously feared the companies' power began to assess it in a more sober light. This tells only part of the story however. There is no doubt that two other major contributors to governments' more relaxed attitude are: (a) their greater experience in handling foreign investments, partly resulting from the activity of international aid institutions, and (b) a world-wide shift in political-economic culture in the 1970s and 1980s towards liberalization and market philosophies, compared to the economic development culture of the 1950s and 1960s which placed great emphasis on planning and intervention in both internal and external markets.

We should not infer however from these observed changes in attitude that governments have lost all their inhibitions about foreign investment, that their confidence is now complete, or that nationalist political instincts have lost all their edge. This is certainly not the case. Governments in developing countries are still jealous of their sovereignty and have clearly perceived goals of national independence. They do not assess the involvement of foreign investors in their countries exclusively in economic terms, so the relationship is not akin to a pure commercial partnership between, say, two private firms. In many cases (though not everywhere) governments worry about the degree of foreign dependence which arises from a large involvement in their economy. Even when the government of the day has no direct cause for concern it may still worry about criticisms from opposition political parties, or about more generalized feelings in the populace related to the resurgence of nationalist and fundamentalist movements.

Lack of Knowledge. Even if they are in general more relaxed now towards the companies, many governments still feel that they are at a bargaining disadvantage in certain areas because of their smaller amount of information and experience. This has particular relevance for gas where the amount of knowledge and skills outside the oil/gas companies is very limited. This unease certainly exists. One of us overheard a long conversation between a top official and his deputy, in one country we visited, in which the former advised his deputy to be extremely cautious in a forthcoming meeting with a delegation from a major oil company. The leitmotiv of his warning was 'they know so much more than we do; who are we in relation to them in terms of expertise; they can easily take us for a ride; they are very powerful in this field'. We shall develop the implications of this in the next section.

To sum up, governments have to cope with a wider set of objectives than a company and face the difficult task of reconciling them. A company is essentially motivated by the opportunities for profits encashable within a reasonable period of time subject to considerations of risk and strategic diversification. A company rarely needs to take any other considerations into account, other than legal requirements. The few exceptions are cases of discreet political pressure by the government of the mother country or pressure from such lobbies as anti-apartheid groups etc. But even in these cases the behaviour of companies will depend on their assessments of future costs and benefits (i.e. ultimately of the impact on future profits) of responding positively or ignoring these pressures. Their concern to maintain a certain public image or to placate the authorities of a country in which the company has major interests, is in the final analysis a function of the importance of that image or good relationship with the government for the 'business' (profits) of the firm.

In contrast, governments always have to consider the trade-off between political and economic objectives (which are themselves wider than those of companies), and these often pull in opposite directions. Political objectives have their own autonomy and unfortunately usually involve economic costs rather than benefits. But their existence cannot be ignored by companies hoping to reach a business agreement with a government.

(d) Financial and Manpower Constraints on Governments. Developing countries suffer from serious constraints on both manpower and finance.

Their most serious lack of manpower is at the level of senior policy-making and negotiating positions. The skills and expertise needed to fully understand the gas industry are very concentrated within the oil/gas companies, and governments recognize their relative disadvantage in these areas, as mentioned above. The consequence is that they are very hesitant to move fast in negotiations with companies, lest they concede too much at too early a stage. This can effectively paralyse all progress, and is very frustrating to the companies. There are steps that can be taken both by companies and aid agencies to alleviate this problem. Already some of the utility companies in Europe offer training programmes, but these could well reach a far wider audience if funding were more easily available. It would be especially beneficial to concentrate on training in the planning and management of gas systems, as well as the technical aspects of the industry, for these would contribute most to the country's capacity to negotiate with companies on an equal footing. It is also important that the organizations providing training do not impart a bias on policy options. Of course the problems arising from the lack of skilled manpower are especially acute in the early days of a country's gas development. Governments, or more precisely their specialized agencies, will acquire knowledge and skills over time.

Finance is clearly a major constraint in most developing countries, and is particularly relevant to gas projects because the government is normally responsible for providing gas infrastructure in the form of pipelines. This level of investment is usually beyond the means of the country itself. An attempt to finance a pipeline in Argentina through a consortium of private investors has recently met with severe problems. While this was largely due to political circumstances it has been enough to discourage similar attempts elsewhere, at least in the short-to-medium term. Given the overall debt situation of developing countries the prospect of commercial banks being prevailed upon to invest in gas projects seems extremely distant, with the exception of a very few countries. Our interviews with banks produced a strongly negative response towards such investments.

The countries therefore need to seek finance for gas infrastructure from the aid agencies. Indeed such investments fall squarely within the original remit of institutions such as the World Bank. Additional advantages also accrue to the country if it can obtain aid finance for infrastructure. First, such a loan will not worsen the country's debt structure; and secondly, companies investing upstream are greatly reassured by the presence of a third party such as the World Bank at the heart of a project. This can make all the difference to their willingness to become involved.

The country's shortage of cash may seem to parallel that of the companies. But the difference is that where companies are likely to recover from their current cash squeeze, many developing countries are likely to continue to face problems for a long time to come.

2.4 The Relationship between Governments and Companies

We have tried in the preceding sections to describe the objectives of companies on the one hand and governments on the other towards gas development, and the constraints under which they operate. We have also compared the complex and frustrating process of government decision-taking with that of companies. It may seem from these descriptions as if the two main partners in gas development are so fundamentally different in their constitution and aims as to raise doubts over the very possibility of reciprocal understanding. We are not suggesting that either company or government will be surprised by our characterization of their own behaviour. It can be difficult however for each party to retain a clear and balanced judgement at all times about the other's behaviour.

Yet one of our strong conclusions is that unless governments appreciate more fully how difficult gas is for the multinational company, and unless companies take very seriously the task of understanding the political economy of each host country, gas projects are very unlikely to go ahead.

To recap. Gas for the internal market is difficult for foreign companies because it confronts them with a greater dependence on and vulnerability to the state, over a larger number of issues and over a longer period of time, than is experienced for any other form of direct foreign investment. In no other case is there a total dependence on a rigid market structure, where the distribution network and entire transportation system are typically (although in principle not necessarily) owned by the state or a national enterprise. This physical infrastructure is extraordinarily inflexible. Furthermore oil companies are rarely given the power to create a market for the gas (a difficult task for them to perform) or to enter into third-party sales. The company is thus directly dependent on state decisions and policies for every element which determines the profitability of its investment, and in particular for pricing, for the volume of sales, and their rate of growth. All these factors are in addition to those which affect foreign investors in all fields, such as tax regimes, legislation, convertibility and the like. In gas, as compared to oil or other resource industries, to manufacturing or even to lumpy infrastructural projects, we are dealing with the long-term supply of a product to specific and usually state-controlled markets over a considerable period of time.[8] The dependence on the state is therefore *continued*.

These features mean that in principle at least multinationals face

---

[8] In oil the output flow is not tied to specific markets; in other resources there are possibilities for vertical integration and some flexibility in market and distribution; in manufacturing production for the local market there is flexibility of outlets and the possibility of capturing *privately* larger market shares; in infrastructure projects the dependence on the state is on a once-for-all basis.

more, larger and different kinds of risks, over longer periods of time, with gas than with other ventures. This is why they demand more favourable terms in agreements with governments for gas than for oil. Higher risks may raise the threshold at which investment is acceptable; or alternatively they may be dealt with through strong or unusual guarantees, either within agreements or through the involvement of third parties. A fundamental part of the risk however arises from a lack of trust between the company and government, especially over long-term behaviour. If this level of trust can be improved then the risks associated with gas can be significantly reduced, although not eliminated. While such trust is a very intangible notion it can be seen to exist in certain cases; one may quote Amoco in Egypt, or Burmah Oil in Pakistan.

Gas also presents difficult problems for governments. There is often a need for them to show greater understanding of the companies' difficulties, but equally the companies need to fully appreciate the political economy of gas. The more important aspects of this political economy have been discussed in the previous section: the macroeconomic impact of the financial commitments involved in building gas infrastructure; the income distributional effects of pricing; the potential conflict with producers or users of other energy sources within the public sector; the significance of political objectives relating to balance betwen regions or socio-economic groups; and the degree of foreign dependence that the state is willing to contemplate.

Thus the country also perceives risk, over a very wide area. The government needs to be reassured about these effects, just as the company needs protection against the risks which it perceives. This calls not only for a better understanding of these issues by the companies, but for serious attempts by them to assess jointly with the government the nature of side-effects, and their real (as opposed to their imagined) significance. Wherever possible remedies are also needed. It is true that companies are often ill equipped for this task. But assistance can be obtained from trusted third parties. The company should also recognize the legitimate fear of the state that it may be driven into conceding too large a share of project profits to the investor, a fear not only of paying out more than necessary but also of the political fall-out if their actions are interpreted in this way. The company must therefore put its case in terms of recognizable characteristics of gas development – economic uncertainty, rigidity, market potential – rather than relying on subjective risk evaluation. An important initial compromise would be for the company to accept that the government is entitled to a proportionately larger share of the reward if the project performs much better than anticipated, in exchange for a promise of help if, through nobody's fault, the project turns sour for the company.

It is important not to gloss over the fact that there is a structural asymmetry in the relationship between government and company. The government is sovereign and the company is not. The government itself enacts the laws which bind it, and can therefore in theory undo any commitments entered into. While these strictures apply to all governments, it is nevertheless a fact of life that foreign companies (and foreign citizens) tend to complain more about the actions and behaviour of host countries than about their own governments. Nevertheless the instances in which governments put themselves entirely outside the framework of international law are very rare indeed, and companies are well aware of this. (Witness the small number of cases referred to international arbitration from countries such as Libya or Iran, which are thought to be neglectful of the international order.) The real concern of companies is not so much that governments may renege on their commitments in an ultimate exercise of sovereignty; rather it is the power of the state gradually to erode agreements, through changes in policy and regulations, new bureaucratic practices, or simply through administrative harassment.

The asymmetry also works in the companies' favour however. The company is expert in the gas business upstream; it has a near monopoly of knowledge and information, including the extent of the host country's gas resources.[9] When combined with its single purpose this gives it considerable power. For the government and its national oil company are, at least to begin with, rarely expert in gas; they have poor access to information; and their attention is diverted by a variety of objectives and a whole host of outside considerations. Companies often use their power in this area by withholding information at critical stages of the negotiation. The excuse given is the need for confidentiality; but the real motive is the hope of gaining a tactical advantage. We believe that in many instances this attitude is counterproductive. It fosters mistrust, delays negotiations and may in the end deny to all parties the benefits of a successful outcome.

Furthermore companies, at least in principle, can invest and operate in any country; they have the whole world at their disposal, whereas governments are restricted to the resources within their own territory.[10] Companies thus have an initial bargaining advantage, since they can seek out the best opportunities available. This may be

---

[9] Oil companies have little experience of the downstream side of gas, particularly distribution and retail.

[10] Very few national oil companies in developing countries have invested in oil and gas outside their territories. Kuwait and Venezuela are recent exceptions.

partly countered by competition between companies, but this is unlikely to play an important role for gas in the Third World. In short
each side wields power of a certain type, and each side suffers from a
corresponding weakness.

This structural asymmetry between companies and governments
lies at the root of many problems; but often manifests itself in dis-
guised ways. Consider negotiations for example. All those involved
are familiar with the difficulties and misunderstandings which arise
sometimes, and which are variously attributed to failure of clear
communication, hostile attitudes, ingrained prejudices and the like.
These factors should not of course be belittled and they may well be
related to deeper causes. But when everything is distilled, the irre-
ducible element is this asymmetry of power. The government's
perception that companies 'know everything' and conceal vital facts
from them ultimately explains obstructionist tactics of government
negotiators. And the companies' perception that governments, *per se*,
will not fulfil their commitments or will be unable to resist the
temptation to vary agreements in response to changing political circumstances, explains the mistrust of companies. For obvious reasons
companies find it difficult to air these concerns about 'political' risks
in open discussions with the government. They cannot say in so many
words that they lack the necessary trust, and even if they could,
governments would not accept it. Hence the various manifestations of
this concealed mistrust in the language, conduct and tactics of company negotiators. A well-known law of psychological behaviour is that
the breakdown of a relationship between two individuals is rarely due
to the disputes or disagreements which both perceive as being the
cause. Rather the real cause is some other grievance or incompati-
bility which is transferred on to an irrelevant aspect of the relationship.
The analogy may well apply to the psycho-drama of negotiations.

Mistrust can therefore cause governments and companies to behave
in unhelpful ways during the negotiations for a gas project. Governments will naturally wonder about companies' behaviour, and ask
why gas projects in their countries are put a long way down the list of
company priorities, or why an apparently unreasonable rate of return
is asked for. This can be easily interpreted as greed on the companies'
part. Similarly companies may misinterpret government hesitation
and delays in decision-making as signs of incompetence or of a lack of
firm purpose, although these are often due to worry about the companies' 'monopoly' on expert knowledge or the real or imputed political intentions of the companies. These false perceptions can produce
an unnecessarily negative atmosphere and are detrimental to the
negotiations.

Two types of rhetoric can further obscure the negotiations. On the one hand companies are sometimes over-anxious to stress their public spiritedness, and their interest in the country's welfare, although their main objective is always to achieve good economic returns on the business in which they are engaged. On the other hand governments, when negotiating with international companies, often go out of their way to stress that foreign investors are welcome in their countries, despite the fact that many of these governments worry about the 'political' intentions of companies, fearing that they may gain a political foothold in the country on behalf of their parent government. The rhetoric of both companies and governments is unhelpful because the truth, sooner or later, reveals itself. A company which claims to be moved by public spiritedness loses credibility in the negotiations when some features of the contract or project turn out to be inimical to the public interest. And a government which claims an unconditional welcome to foreign investors will create ill feeling when it tries to impose tiresome rules and regulations on the operations of the international company because of its unspoken fear of the company's political intentions. Ultimately it is preferable for both sides to be explicit about their true objectives and to be candid about their important worries.

A further aspect of the company/government relationship is usually hidden behind a thick curtain of silence, but needs to be unveiled. The phenomenon goes under different names: 'corruption', 'commission', 'speed money', to name but a few. Propriety and feasibility have prevented us from studying the problem directly. But we were urged, and very emphatically by one senior civil servant from a developing country, not to become implicit participants in the conspiracy of silence.

Our views on the subject are not based on specific instances encountered in the country visits, or on revelations made by interviewees, whether from companies, banks or international organizations. They rely rather on a broad knowledge acquired from studies of economic development and business behaviour. We should also emphasize that the problem is far from being peculiar to either gas projects or to the Third World. Corruption is unfortunately a widespread phenomenon which affects several areas of business activities and exists in different forms all over the world.

We define the problem in terms of payments, direct or indirect, from one party to a person on the other side of the divide, whose function is to make decisions about a project or a transaction. This may occur at any stage of negotiation, finalization of agreements, or implementation. The phenomenon has been described by some social scientists as one of income redistribution. Pseudo-moral justifications are quickly found, particularly when 'speed money' is demanded by and paid to low-ranking and very badly paid officials, who have power of obstruction on a wide variety of issues and can cause the firm or the contractor losses of time or money. Small gifts seem to oil the cogs of the machine and speed implementation, and at the same time correct some inherent inequity in the distribution of income.

Yet the same justification is also sometimes offered for the actions of higher-ranking officials, on the grounds that their decisions, however properly determined, nevertheless provide the business concern with a large profit income which would not otherwise accrue to it. The practice is then expressed in terms of 'services' provided and 'commission' paid in return. The language transforms the real meaning of the phenomenon however, since services and commissions are morally neutral concepts in business and economics. Some even go so far as to suggest that things are better than neutral: 'it is only fair to recognize the help received', 'it is only equitable (read, it would be morally wrong to act otherwise) that we help you when you help us'. These statements are not pure invention: they are often uttered and heard.

The fallacy is that in this context payments to individuals simply redistribute income. In fact, all payments and commissions paid to officialdom for the purpose of a business gain are likely to involve costs for the country. They cause decisions to be biased against the best tenderer in favour of a second-best, concessions to be made that reduce the benefit of the investment or transaction to the country, inflated prices to be paid, sub-standard execution of work, etc.

Another fallacy is that foreign firms cannot make such payments because of legislation in the mother country. Bribes need not be paid as bribes, and firms can make payments indirectly without falling foul of the law. One method is to hire a local consultant and pay him a handsome fee, which he then channels (in part) to an official. Another method is to engage a local partner, which in any case is mandatory in some countries. The local partner retains part of the profit, or charges the joint venture for labour and services that are never fully provided, and then shares the income with an associated group of officials. A third method is for the foreign firm to buy an asset, say an office flat, at an inflated price from the associate of an official and then resell it after a while at a lower price to another associate of the same official. Examples can be multiplied.

Finally, there is a fallacy which would have us believe that, whatever their drawbacks, payments and commissions at least speed things up. This may seem to be true in the daily interface of a private individual with low-ranking officials: the foreman, the customs officer, the policeman, or the bureaucrat who can sit indefinitely on an application for a permit or an invoice. The truth however is that the essence of corruption is to create a delay where there is no need for one, in order to create opportunities for personal gain. For example, a government may require companies to ask for permission each time they want to move an exploration rig. At higher levels of decision-taking the system of 'services and commission' certainly increases both frustration and delays. The reason is that at this higher level no official has sole and exclusive discretion. Commissions have to be shared by a group; and since mutual trust between those concerned is critical, officialdom usually splits up into small and rival groups. There is competition and struggle which may snare the unwary in a fatal web. The rivalry also causes delays because each group has to move very cautiously, either to ensure that its actions do not elicit damaging retaliation from another group, or to negotiate a sharing of the commission with rivals. In this game it is essential to create the appearance that everything is above board. The additional delays are therefore justified in all sorts of ways: the matter in hand must be carefully evaluated over and over again; there is endless haggling in negotiations over minor items or clauses; the fear of making concessions is exaggerated in some instances to detract attention from the major accommodation. All these procedures waste a large amount of time and end up by costing large sums of money.

To conclude. When companies and contractors acquiesce in corruption in order to do business it is a cause of delays, resource misallocation and additional costs which are inevitably borne by the project (therefore by the country). On the other hand, when companies and contractors are unwilling to get involved with illegal practices the existence of corruption may deter them from becoming involved in projects. In both cases, albeit in different ways, gas development suffers.

2.5 Conclusion

It should not be inferred from this whole discussion that successful relationships between companies and governments are impossible. The partners do have different identities, different aims and a different balance of strengths and weaknesses, but both governments and companies are surprisingly adaptable. The problems we have discussed can be overcome, providing the following conditions are met. First, there must be a mutuality of interests, in other words some significant gain possible for each party. Secondly, they must be willing to work out an agreement which is both *flexible* in its detailed clauses and *secure* in its basis. This must stem from an equitable distribution of benefits between the partners and a reliance on their mutuality of interests, rather than on subjective trust, for its survival. We shall turn to the practical and policy aspects of these issues in the next two chapters.

3 THE PROBLEMS OF VOLUME, PRICE AND RENT

3.1 Introduction

When considering a potential gas project three major issues dominate the appraisal and planning of the project; the negotiations between governments and companies for both exploration/production agreements and sales contracts; and indeed their whole relationship. These are the issues of volume, price, and rent.

There are two major sources of uncertainty about the volume of gas in a proposed development. The first relates to the likely market outlets for gas, the volumes that can be absorbed by these markets and their future rates of growth; the second relates to the size of current and future discoveries of gas. These problems have a compound impact on the gas project because gas production is usually tied to specific outlets. Unanticipated imbalance between supply and demand cannot be easily corrected, and can result in either disruptions at the market end or considerable losses for the producer (or both). The only way to cope with these uncertainties is by careful planning of the gas development with due attention paid to the interdependence between upstream, downstream and transmission, and the need to synchronize different stages of development in these three areas (this issue is discussed in Chapter 4).

On prices, the fundamental issues are the concepts and principles which should govern the setting of producer and consumer gas prices, and the setting of pipeline tariffs. There is also the problem of uncertainty over both future international energy prices and future pricing policies of the host government. Finally there is a complex relationship between pricing and volume, largely because volume has an impact on costs. Early agreement on clear pricing principles for gas at the well-head can considerably improve the chances of a potentially good project by contributing to speedy negotiation of contracts. Similarly, well-designed pricing policies for consumers are essential for the promotion and establishment of markets for gas and hence for the prospects of a projected development.

Finally, complex issues arise as regards the sharing of natural resource rent between the government and companies. Governments face the task of designing fiscal regimes which provide companies with just the right incentive to undertake gas projects that are of benefit to the country. Ideally the government should neither give away too little, as this is self-defeating, nor too much, as this is irrational. For a wide variety of reasons the balance is difficult to strike. First, there is much uncertainty over volumes, prices and costs, making it virtually impossible to guess in advance the *actual* distribution of income resulting from a particular tax system. Many fiscal regimes, for example, specify tax rates which are a percentage of some fiscal concept of income; but rates which appear to be reasonable at the start of a gas development may in practice yield too much revenue for the producing company if the fields turn out to be much bigger or cheaper to operate than initially expected, or if there are unanticipated increases in world energy prices. Conversely they may penalize the company too much and kill the incentive for further development if costs turn out to be higher, prices to be lower or volumes smaller than expected. Secondly, it is not always easy for governments to assess the minimum level of fiscal incentive which producing companies will effectively accept; this may take lengthy negotiations, much bargaining experience, and sometimes costly trial and error. This is a familiar bargaining problem. It plays an important role in gas as in oil, because the stakes are high for both companies and governments; but it is a trickier problem in gas than in oil because there is less experience on international gas than on oil, and a more limited body of reference cases to draw from. Thirdly, there are also technical difficulties in designing a fiscal regime which achieves both the desired distributional objectives and the variety of allocational objectives which governments try to achieve with tax instruments. Ideally taxation should be neutral from an allocation standpoint, but this cannot normally be obtained without undesirable distributional effects.

3.2 Volume

(*a*) *Market Uncertainty*. There are *potentially* three main outlets for gas, namely: export markets (both LNG and exportable gas products such as methanol or fertilizers); direct arm's length sales to a local major end-user; and sales to the transmission intermediary (in developing countries usually a state enterprise), which then resells the gas to the end-users. In all these cases, albeit for different reasons, there are significant uncertainties.

In this study we are not concerned directly with LNG projects; but nevertheless the issue has relevance in this context. In normal circumstances a foreign company is more attracted to countries where the gas potential could result in an export project than to countries where the volumes of gas are likely to restrict sales to domestic markets only. Obviously the main reason is that exports generate foreign exchange earnings, thus removing the thorny problem of convertibility. Exports also enable the company to benefit from upward movements in international prices, an opportunity that may not exist in domestic markets where prices are either fixed or tightly regulated by governments.

Although the export prospects for LNG are rather dim today a future improvement, probably in the late 1990s or early 2000s, cannot be entirely ruled out. All other things being equal, companies remain more interested in countries with large potential gas reserves (necessary to support LNG projects) which are also close to big export markets than in countries with small reserves, or countries that are not favourably located in relation to Western Europe or Japan.

In many countries, however, gas reserves are not large enough to ensure that export commitments over the long life of a contract will remain consistent with future (presumably growing) requirements of the domestic economy. In these situations uncertainties about both the reserve potential and future local demand make governments unwilling to contemplate exports. This unwillingness is perfectly understandable because it is not easy to import gas, should a supply shortage emerge in the local market. There is however a disincentive to companies when governments feel unable to dangle the carrot of export possibilities in the foreseeable future. This naturally affects the chances of an earlier or complementary development of gas for the domestic market.

Governments sometimes attempt to solve this problem by introducing the concept of a national reserve. The idea is that exports will be allowed once aggregate discoveries build up to a volume of reserves deemed adequate for domestic requirements over a certain time-horizon. Canada follows a complex procedure to determine the size of the exportable surplus, defined as the amounts which may be disposed of out of reserves in excess of local needs. Egypt is one of the countries covered by this study which adopted the concept of a national reserve, although this has now been abandoned. We are not fully convinced that such arrangements provide a strong incentive for exploration. They involve of course the promise that exports will be allowed when the national reserve reaches its target level. And, as mentioned earlier, this promise had some attraction in the recent past and may become valuable again some time in the future. The drawback for companies, however, is that if they explore for gas just to add to the national reserve they are unable to capitalize immediately on their discoveries, unless there is already an existing domestic outlet, together with suitable arrangements governing such developments. In Egypt, the concept of the national reserve was introduced before these necessary arrangements. The government proposed that the state acquire the reserves against compensation for the exploration costs incurred (plus interest). This policy did not encourage exploration specifically targeted at gas-prone areas, and merely provided a bonus to companies which happened to find gas while searching for oil.

In the majority of cases, particularly in today's conditions, the internal market offers the only immediate possibility for developing a gas resource. An internal market can be served either by public enterprises or by the producing companies diversifying downstream.[1] Where possible, companies are attracted by third-party sales to the electricity generating sector or to large industrial customers. They naturally prefer such outlets to dealing with monopsonist state enterprises or regulated public utilities. However, most governments do not appear very keen to encourage this option. They are right, but for the wrong reasons. Usually the government's reluctance to allow third-party sales is due to the self-interest of government agencies and state enterprises. The real problem of direct supplies is different however. They may preclude small customers and might take volume to such an extent that a separate distribution scheme for them becomes infeasible. They may thus prevent the planning of a good transmission system with common carrier features which could enable future diversification of customers and uses.

Where the outlet for gas is the internal market and where this market is served by parastatal enterprises the gas producer will usually worry about such a bilateral relationship with a monopsonistic buyer. Leaving aside for the moment the question of confidence in the buyer's continued commitment to the pricing principles initially agreed upon, the producer may be legitimately concerned about the security of the volumes guaranteed by the parastatal. Usually the guarantee takes the form of take-or-pay clauses in the gas sales contract, but experience has abundantly revealed (as evidenced by recent events in the USA) that such clauses do not provide much protection when market circumstances change in significant and unanticipated ways. A take-or-pay clause is essentially an initial expression of goodwill and a promise of good behaviour, not an insurance against *force majeure*. The only real guarantee lies in an informed conviction that the market for gas has been well appraised and

---

[1] Some companies are even prepared to contemplate direct involvement downstream, including the distribution infrastructure and the promotion of a market in the residential/commercial/small industry sectors. (So we were told by some, perhaps exceptionally forward-looking, company planners with an interest in Nigeria.)

properly promoted. This is precisely the reason why the planning of the gas project as an integrated and synchronized package of upstream, transmission and downstream components is so important, as argued in Chapter 4.

The purpose of this planning is to cope with (not to eliminate) the many uncertainties on the demand side of a gas project.

Examples of such uncertainties are many. The construction of power stations and industrial plants, initially expected to contribute to gas demand, may be delayed or even cancelled because of changing economic conditions. This has happened in Egypt, Nigeria and Thailand to name but three examples. While the development of a gas field is in progress a firm lined up to become a major user may change its mind and shift to another fuel before the sales contract is finalized. Customers may also be lost, or permanently reduce their demand if the gas development runs into teething troubles. This happened in Thailand when the Unocal field failed to deliver the quantities contracted for. In a dynamic perspective there are also major uncertainties about the growth in demand for gas because expansion depends: (a) on investments in the transmission infrastructure, which may be affected by a deterioration of economic conditions or of the debt situation; (b) on pricing policies which are liable to change; and (c) on the success, difficult to predict, of the market promotion effort.

(b) Uncertainty about Reserves and Future Discoveries. Uncertainty about reserves and future discoveries has more significant implications for gas than for oil projects. The lumpiness of gas transmission systems means that construction of a pipeline will not be economical until the supporting gas reserves reach a minimum size. Gas may be initially discovered in smaller quantities than this critical minimum size. The company will then have to wait before it can develop the reserve, causing a lower return on exploration investment. This may be exacerbated if it has no opportunities for further exploration, say for lack of concession acreage. Such situations call for more generous stipulations on the allowable time-lag between discovery and development in the exploration and production agreement.

The reverse situation can also be a problem. Because of the limited size of the local market a situation may arise in which one large discovery fulfils the the whole market requirement, thus pre-empting for a while the development of subsequent discoveries. In a sense this is a risk the companies must accept. They can only console themselves with the idea that a field discovered too late is akin to a dry hole. But this possibility should caution governments about the wisdom of encouraging too much gas exploration (usually at the cost of bigger incentives than otherwise necessary) when the potential market is small.

Finally, there is some uncertainty attached to associated gas, simply because it is a by-product of oil with its output rate dependent on conditions governing the production of oil and the behaviour of the field. This has to be taken into account in planning for the use of associated gas.

3.3 Prices

(a) *Price Concepts and Pricing Principles*. An important distinction must be drawn between four price concepts for gas:

(a) the price which expresses in opportunity terms the gross value of gas to the economy, say P_r;
(b) the net opportunity price which expresses the net value of gas to the economy, say P_v;
(c) the price P_c charged to end-users;
(d) the price P_p paid to producers.

Opportunity Prices. One important aspect of gas pricing is the need for a reference for meaningful economic valuation of the resource. Economists cogently argue that the border price of a good or commodity is both a convenient and theoretically defensible measure of its opportunity value. This equivalence applies when the commodity is in principle tradable; it need not actually be traded. There is no price standard for gas in the international market, as LNG is sold in bilateral trade arrangements under price formulae which vary between contracts. Gas is however tradable in the sense that it is a substitute in most uses for oil products which are extensively traded. (In some cases, though to a much lesser extent, it is a substitute for coal which is also traded, but the price of coal is usually set with reference to fuel oil.) This suggests that the international price of oil can provide a reference to the value of non-traded gas in domestic use. There are complications, however. Crude oil is not the correct reference because gas is a substitute for specific products: fuel oil in steam raising, gas oil in space heating, diesel oil in peak-load electricity generation and certain industrial uses, LPG and kerosine in the residential and commercial sectors. Gas is also a feedstock, and in this area the substitutability with oil is imperfect; oil products and gas are used for the production of different petrochemicals. In short, gas is a good substitute for some oil products as a fuel (the major exception being certain transport uses, where gas substitution can entail high costs) and an imperfect substitute for the manufacture of certain petrochemicals. To take the crude oil price as a reference will be misleading on two counts: (a) the imperfect substitutability that has already been explained and (b) the lack of correspondence between the pattern of gas use and the composition of the refined barrel of crude taken as a reference. Of course adjustments can be made for these two major differences but the exercise is not very straight-forward.

In short, the price P_r which expresses the gross value of gas to the economy in opportunity terms is the border price of the relevant substitute. There are practical difficulties in identifying the fuel (or in constructing the basket of fuels) for which gas is the actual substitute. But the basic notion is straightforward: a btu of domestically produced gas saves a btu of imported fuel, or releases a btu of fuel for export. Thus the unit value of the imported/exported fuel (converted and adjusted for the purposes of correct reference) is the opportunity value, in gross terms, of gas to the economy.

However the development of gas reserves, the extraction of gas and its transportation involve costs for the economy. These are:

(a) the opportunity cost of the domestic resources used up in development, extraction, etc.
(b) the sums paid to foreign companies (producers, contractors, etc.) for inputs of foreign origin, including their share of the mineral rent and any monopoly rent which they may acquire.

The net value of a gas development to the economy is therefore the difference between the value of the substitutes and these costs measured in opportunity terms. But this is not the end of the story, as it is necessary to take two other factors into account. First, gas development and the replacement of other fuels by gas in the energy consumption mix involve externalities, mainly but not exclusively environmental. There are social benefits from using gas, for example a reduction in the pollution levels in towns, and in some cases social costs, for example higher accident risks. In assessing the net value of gas to the economy it is important therefore to add the value of external benefits and deduct external costs. Secondly, since gas is an exhaustible resource, the problem of optimal depletion arises. This problem can be ignored where the reserves are very large, but could be important in countries where they are known to be limited in quantity and liable to be exhausted over a relatively short period of years. There are opportunity costs to the economy whenever the time pattern of gas production diverges from the optimal depletion path. Theoretically these costs should be included when assessing the opportunity value of gas development for an economy. In practice however, this aspect is generally ignored. One reason is that governments, except in special circumstances, do not worry about the long-term aspects of the production of an exhaustible resource.[2] They are generally more concerned with immediate benefits and revenues than with hypothetical gains from conserving the natural resource for future exploitation in times of higher prices. Furthermore it is not easy to work out a meaningful solution to the depletion problem because uncertainty over future energy prices and actual size of reserves is too great and the problems of working out an optimal depletion path are too complex.

To sum up. The net value of gas to an economy is the value of its substitute plus net external benefits minus all relevant opportunity costs. This concept is very important because it indicates whether a country should or should not promote a particular gas project. It also helps in determining the size of the 'value pie' which can be shared by the various participants.

Consumer Prices. The reference for consumer prices is the price paid by end-users for substitute fuel. If gas is to be adopted at all by an end-user P_c must initially be lower than the price of the alternative fuel, in order to provide an incentive for a switch-over and to compensate the new customer for the capital cost and the trouble involved in converting a fuel-using appliance or plant.

Furthermore, in consumer pricing it may be advisable to give end-users a greater incentive to convert to gas than the minimum discount required to compensate them more or less exactly for the direct costs of converting. Such a policy may promote greater demand for gas and yield benefits in the form of cost reduction through economies of scale. When there are increasing returns to scale, the price of substitutes is not a sufficient guide in itself to market-related pricing.

There is scope for price discrimination between different users of gas, because transfers between say a power plant and a household are not easy. Thus P_c can be set with reference to the price of fuel oil for a power station which will use gas only for base load, and to the price of kerosine or LPG for the residential sector.

In most countries the prices paid by consumers for fuels diverge widely from their border (or opportunity) prices. This divergence is

---

[2] The notable exceptions are the USA in the 1920s and 1930s, when worries about wasteful oil production led to strict output programming, and Kuwait and Libya in the late 1960s and early 1970s, when concern about the fast depletion of their oil led them to impose conservation measures.

due to taxes and subsidies, to price regulation and to various economic distortions. In many Third World countries the effect is significant, with energy pricing policy tending to keep consumer prices for fuels well below opportunity values. Such policies induce a greater demand for energy than would obtain if consumers were charged the (higher) opportunity price. This effect may be beneficial for gas development if there are economies of scale and the larger demand creates the critical minimum size for a project to take off. There are instances, however, where the need for gas to compete with other fuels which are priced uneconomically on the internal market is a deterrent for gas projects. The obvious case is where low consumer prices are achieved by fixing the producer price at a level which does not provide the producing company with acceptable returns. In other instances, low consumer prices are achieved through direct or indirect subsidies which need not distort producer prices. In theory a consumer subsidy which does not interfere with the price paid to producers or with transmission tariffs should not have adverse effects on a gas project. It may however inhibit development if companies fear that the government will be compelled to reduce the subsidies in the future, and will then attempt to mitigate the effects of this on consumers by interfering with producer prices.

Producer Prices and Transmission Tariffs. Producer prices for gas have historically been set in either of two ways: market-related or cost-plus.

One approach is to pay producers a *market-related price*, which is essentially a netback to the well-head of the border price of a substitute fuel. In many instances (e.g. Egypt in the new model gas clause) the reference is the international price of fuel oil. A case can be made however for a different reference: a weighted average of the border prices of substitute fuels, the weights being the proportions of gas used in different sectors according to the fuel displaced. Thus the weight of fuel oil, or in some countries of coal, will correspond to the share in gas use of power and industry (steam-raising use), that of kerosine or LPG to the share of the residential sector etc. A minor complication is that the shares of different sectors in gas use will vary over time and will thus require adjustment of the weights at regular intervals. This could be done perhaps annually on the basis of the average shares of gas use in the preceding year.

The producer price set through a market-related approach is a netback, and thus involves the deduction of cost elements from the point of use back to the well-head. The value of the method (and hence the price paid to producers) is sensitive to the cost concepts used and to the methods applied in measuring costs. We shall not however discuss in detail the problems which may arise in defining or computing cost elements, other than transmission costs. The latter are often significant and have a heavy weight in netback calculations; they therefore deserve to be singled out for discussion.

Transmission costs depend, among other things, on the method chosen for setting pipeline tariffs. Some methods make the costs appear to be much larger than do others. They thus yield lower netbacks and lower producer prices, and unnecessarily discourage gas projects. But how should transmission tariffs be set?

It seems reasonable, at first sight, to set transmission tariffs in relation to unit costs. But a number of problems immediately arise. Unit costs depend, other things being equal, on the volume of gas transported by the pipeline. This volume varies over time: first because there is an initial build-up period for both production and demand, secondly because of subsequent growth, and thirdly because of unavoidable fluctuations. Thus, unit costs are likely to be higher in the initial years than later on. If tariffs are set strictly in terms of unit costs they will be high in the first years and will then become gradually lower. This may seem logical but it puts an undue burden on the producer in the early years of production when cash flow needs are higher because output is low. A fairer approach to both producer and owner of the transmission system (the costs incurred by the latter are largely in the nature of fixed costs and do not vary very much with the volume of the gas throughput) is to set the tariff on the basis of average unit costs over the build-up period. This leaves us with the issues of subsequent growth and temporary volume fluctuations. Growth benefits all the parties concerned and can be accommodated by reductions in tariffs when permanent growth occurs (if for some reason there is a permanent decline in volumes, then tariffs have to be raised accordingly). The tariff level can take into account anticipated fluctuations, allowing for best estimates of their likely range of throughput. Unanticipated fluctuations call for temporary adjustment of tariffs. Another difficulty arises when the pipeline is built to a very large size in anticipation of volume growth in future years. In such an instance there is a prima-facie case for a subsidy on the tariff if the pipeline is owned by a government agency (the most likely occurrence in a developing country).

Transmission costs involve a large element of capital expenditure. Other things being equal, the corresponding netback deduction will depend to a large extent on the choice of amortization period for the transmission network. Here again, we believe that where the pipelines belong to the state, transmission costs and other relevant cost items involved in the netback should not be inflated in order to capture rent indirectly. Rent appropriation is better left to fiscal measures.

Finally, a clarification of price concepts may be in order. When a market-related price is applied in the context of a production-sharing agreement (say when the international price of fuel oil is taken as the reference), it is important to distinguish sharply between the 'economic price' and the 'fiscal price'. Confusions on this issue are not uncommon and can cause unnecessary trouble. The problem may be stated as follows. Under production sharing the price paid to the producer for gas applies to the producer's share of the output. The remaining share is acquired free by the government. In some cases, the government may be willing to concede to the producer a price for his share that is higher than the netback of the market reference price of the substitute fuel. This is because there is leeway, given that the average price paid for the whole volume of gas produced is significantly lower than the netback (the government share being at zero price or just at cost of extraction). It is important to realize, however, that when such a concession is made, the price paid to the producer is *not* the economic price. It is rather a price raised above its economic level by a tax rebate. Such a price ceases to constitute a relevant reference for downstream pricing. When this point is not clearly understood the government may seek, as apparently happened on one occasion in Indonesia, to charge this higher price to the national oil company for *all* the gas supplied and the national oil company (or any alternative intermediary) will find it impossible to on-sell the gas charged at such an uneconomic price. This causes unnecessary problems, and probably delays or even arrests development.

The other method for determining the producer price is the *cost-plus approach*. As its name indicates this method relates the price of gas to the costs incurred by the producing company, with an additional element to provide the company with a return on the capital invested. Different concepts of costs may be used in this approach: an *ex ante* estimate of the costs likely to be incurred in the development of the field (Thailand for the Unocal field) or an *ex post* audited account of costs actually incurred (Pakistan for the Sui field). In the first case the base price (which is then adjusted by an escalation formula) includes an imputed element for return which may or may not be exactly realized. In the second case the price includes an agreed *unit* profit margin which is fully realized. The cost-plus pricing method is meant to provide the company with an adequate rate of return on the capital invested to develop the gas field. But the rate of return which a company actually obtains under the cost-plus method is unlikely ever to correspond to the rate negotiated. One main cause of divergence, which arises whether the costs are estimated *ex ante* or *ex post*, is that the volume of gas produced is usually different from the amount initially assumed. Larger (smaller) volumes than anticipated, given the cost-plus price as calculated by the formula, mean a larger (smaller) income for the company, and therefore a different rate of return on the capital invested. The *ex ante* cost approach involves another cause of possible discrepancies, namely the likely divergence between actual costs incurred in a given year and the costs imputed by the method for the same year.[3]

For a producer the difference between a market-related and a cost-plus price for gas is analogous to some extent to the difference between a variable-dividend and a fixed-interest asset for a financial investor. Of course, neither an equity nor a bond provides a fixed return because they are both subject to variations in their prices. Similarly neither the market-related nor the cost-plus method ensures a fixed return on the gas investment. In the former case changes in total returns depend on changes in both volume and price; but in the latter case only on volume changes. An economic agent's preference for one or the other method depends on the degree of risk aversion. The market-related approach to pricing shifts part of the risk of price movements on to the producer (the risk is always shared with the government through taxation). The cost-plus method insulates the producer from the risk of price movements, leaving the government to carry their full impact – obtaining the full increase when prices rise, but losing the amount by which the price declines when there is a fall.

In interviews, most oil companies told us however that they do not like the principle of cost-plus pricing because they are motivated by risk-taking for high returns. They emphasized their preference for the market-related approach. At first sight this stance is difficult to reconcile with many recent manifestations of risk aversion by the oil companies. In fact what the companies are probably trying to say is something rather different. Namely they will not commit themselves to a project if they judge the downward risk to be relatively high and the upside potential to be low. In such a case the guarantee of a unit profit margin in the cost-plus method is irrelevant. They will only commit themselves if the prospects of a good outcome (i.e. a large discovery or an upward movement in oil prices) are relatively high. In such a case the cost-plus method is unattractive because it limits the benefits, and the market-related approach is preferred because it enables companies to capture part of the upside potential.

For governments the cost-plus method seems more attractive than

---

[3] As indicated above, costs in year t are estimated to be equal to the costs as initially estimated × escalation formula.

the market-related approach when a field is large and the unit costs are low, because governments can then extract the maximum possible share of the economic rent through pricing. The same result may not be as easily attained through taxation (which is generally a more appropriate instrument for acquiring the rent), because it will involve taxation rates that are so high as to discourage exploration. But we are told that insistence on cost-plus pricing is a disincentive to companies. These difficulties take us to the heart of the problem: the delicate balance that needs to be achieved between the government's claim on part of the rent and the need to provide incentives to achieve the desired rate of development (see Section 3.4).

(*b*) *Price Uncertainties.* A *company* involved in gas development for the internal market faces two types of price uncertainty. First, there is uncertainty about future oil price movements on world markets. This has a considerable influence on future income from a project, not only when the producer price is market related, but in all circumstances because of demand effects. The nature of uncertainty about oil prices has changed in recent years. In the 1960s the presumption was that the oil price would remain fairly stable and low, allowing for some increase that oil companies were prepared to eventually concede to placate oil-exporting countries. In the 1970s and until 1981–2, the predominant expectation was of future oil price rises. This expectation induced an interest in gas. It also explains certain companies' and governments' positions, which still survive, about such issues as market-related and cost-plus pricing (the conventional wisdom dies hard). Today however the dominant expectation is that the oil price will fluctuate around an average price in the middle teens, with the possibility of sudden but short-lived price collapses (to below $10 per barrel) and an oil price recovery some time in the 1990s. Very recently a growing body of opinion is beginning to have doubts about a price recovery in the early 1990s.

Companies' attitudes towards gas development are strongly influenced by their own price scenarios. Those who believe that the oil price will recover may be prepared to take some risks with gas investments relatively soon, and they will certainly press hard for a market approach to pricing, because of their favourable assessment of the 'upside risk'. Naturally one cannot expect much involvement from those companies with pessimistic views about the long-term oil price. Whether these companies could be motivated by contracts guaranteeing a certain level of return, an approach consistent with their own assessment of risks, is however doubtful. Companies talk about risks in one way and take risks in another.

Secondly, there is uncertainty about changes in domestic energy prices where these are not closely associated with international prices (that is, in most developing countries). The gas producer will also worry both about government policies and their stability, and about monopsonistic behaviour of the state enterprise, the purchaser of gas. There are different aspects to this price uncertainty.

It can arise at a very early stage when loosely drafted gas agreements (or more usually petroleum agreements that failed to properly address the gas issue) are vague over the pricing regime. It is essential to define from the outset the pricing *principles* (but not necessarily all the detailed price clauses) which will govern a gas development. We have already expressed our preference for market-related pricing, but are aware that this approach has its difficulties and may not be appropriate in all circumstances. When adopted, care should be taken to allow for flexibility on: (a) the concept of the reference market price (the changing pattern of gas use changes the relative significance of different substitutes); (b) the cost elements of the netback formula (these should be subject to revision by mutual agreement at regular intervals); (c) the size of the discount required for the promotion of new markets; and (d) the periodicity of the gas price adjustment to the market reference (the option to adjust at shorter intervals than initially provided is helpful for periods of high price volatility).

There is a particular difficulty with the market-orientated approach when the reference is international prices, and when these are higher than the prices charged to consumers domestically. In such a case the subsidy to consumers is likely to appear as a deficit of the state enterprise which purchases the gas from producers for resale to end-users. From an economic point of view, an accounting deficit that corresponds exactly to an explicit subsidy need not pose problems. But governments world-wide, the state enterprises themselves and the international aid organizations, dislike accounting deficits, albeit for different reasons. (Recall the abolition of BNOC in the UK, which was justified on the grounds of 'deficits' which were in fact due to oil policy, not to administrative ineffectiveness.)

One solution is to reform the domestic energy price structure, a goal pursued with success in some developing countries and bitterly resisted in others. It is however extremely difficult for governments to move from a highly subsidized pricing system to a market-related one, in both political and purely practical terms. Two issues need to be addressed. First, there is still a large task to be carried out in persuading governments that such a change is necessary. The message probably cannot be effectively transmitted through the prescriptions of international organizations, which arouse much resentment of interference. The second important issue is the need for practical help in carrying out price reforms, in those countries where governments do decide to undertake them (whether freely or under duress). To simply spell out the ideal end position is far from sufficient. Governments need suggestions and proposals for how to implement the steps towards the goal. Political unrest can be avoided if a set of new fiscal and economic policies aimed at alleviating the hardship imposed by price reforms on low-income groups are implemented together with price changes. Surveys should be carried out to identify the social groups likely to be affected by the price reforms, and effective mechanisms for directly subsidizing poorer sections of the community devised. There are further complications in countries (particularly Latin American) that suffer from inertial inflation. Policies aimed at adjusting relative prices can elicit new inflationary pressures, and prove in the end self-defeating. To correct severe price distortions in countries where these have existed for a very long time is truly a large task. Governments' irritation at the shrill exhortations of USAID officials is understandable. It also seems that the advice given by the World Bank and the IMF, though more thoughtful, often fails to address the social and political complexities of the issue.

Another solution is to adapt the market reference to the domestic price levels, and compensate the producer through a reduction of the fiscal burden. A major drawback is that companies are more prepared to take risks on international prices than on domestic pricing policies. A third solution, which we certainly prefer, is to face squarely the problem of the accounting deficit of the state enterprise by a direct transfer (either in the books or in the coffers) of the subsidy given to consumers. This can be done by earmarking a source of revenue (e.g. gasoline tax) for this purpose. One advantage of this solution is that it turns the implicit subsidies embedded in prices into a visible amount of cash, and this may sharpen governments' perceptions about the real significance of subsidies. The problem, of course, is that governments' attitudes may change over time making direct subsidization unreliable. The subsidies are eventually reduced or removed, and governments then attempt to push producer prices down in order to mitigate the effects of reduced subsidies on consumers.

Thus, while the first step is to define the pricing principles unambiguously from the outset, companies will also need reassurance against future changes in these principles which may be induced by a deterioration of economic circumstances or changes in government policies. In one sense this is a matter of trust; but confidence in the maintenance of agreed pricing principles also depends on the degree of consistency between these principles and the prevailing economic system. Companies may doubt that a government, under the pressure of events, will be able to retain a reference to an international price (for producers) when prices charged to consumers are significantly lower than this reference. Thus, it is not enough to design good pricing principles for gas producers. These principles will lack credibility if the market-related approach is at odds with other major economic policies of the country. There is a serious difficulty here. It is facile to argue that the solution is to run the whole economy on market-oriented lines, as this may not be always possible, and in some cases not necessarily desirable. The realistic approach is to concentrate on the main distortions of the economic system and to try to remedy them gradually but consistently. Such an approach lends more credibility to governments' intentions (as expressed, say, in the adoption of market-related gas pricing principles) than would a comprehensive attempt to open up and liberalize the economy in one fell swoop which ends in failure.

Governments also face the problem of price uncertainty, particularly uncertainty about international energy prices. In the 1970s expectations of ever-increasing prices may have influenced, to some extent at least, Thailand's preference for cost-plus pricing. As things turned out the oil price collapse of 1986 falsified these expectations and put enormous pressure on PTT. The cost-plus price paid to Unocal did not adjust to reflect the lower prices of oil (partly, but not only because of the long time-lags in the adjustment formula), and the end-user price (paid by EGAT) came under pressure through threats to substitute cheap fuel oil or coal for expensive gas. Governments would therefore be well advised to consider sharing the market price risk with companies, since there is no assurance that oil prices will rise significantly for many years. And if this expectation is falsified and prices do rise then let companies get some of the benefit, since it is the dangling of this particular carrot (the upside potential) which enhances their incentive to invest in the gas project. Governments, for obvious political reasons, fear this eventuality. They should however clearly explain to their political constituencies, from the outset, that in a world of widely fluctuating prices the case for risk-sharing is extremely strong.

3.4 Rent

The pricing issue raises the related question of the gas 'rent' or the 'economic surplus' generated by the project, and particularly of the sharing of this surplus by the parties involved in, or benefiting from, the gas project. There is a potential surplus (above normal profit[4]) per unit of output whenever the opportunity price of gas to the economy P_r is higher than the opportunity costs of the factors of production applied to produce and dispose of a unit of gas.

We have argued that a gas project should only be undertaken when the difference between opportunity price and costs is expected to be positive. Thus, by definition, all good projects entail an economic surplus, and the thorny question of entitlement to, and appropriation of, this rent now arises.

Let us emphasize first that the economic surplus generated by a gas project cannot be exclusively attributed to any one part of the project. It is nonsense to say, for example, that 'the surplus emerges upstream because the producer takes the greater risks' or 'because the producer initiates the development'; or equally to say that 'the surplus emerges downstream because production is worthless when there are no consumers'. Gas has economic value only when there is production *and* transport *and* consumption. The economic surplus is attributable to the project as a whole, in other words to the integrated set of upstream, transmission and market development activities. Every part of the project is necessary and no part is by itself sufficient for an economic surplus to emerge.

The co-operation of all the agents involved in a gas project is therefore instrumental in realizing the gains to the economy, but this does not mean that these agents are necessarily entitled either to the whole or even to a part of that surplus. Strictly speaking the country alone is entitled to the surplus as defined above, and it is then up to the government to decide on its disposition.

However, in real world situations, the actual economic surplus generated by a gas project for appropriation by the country is likely to be smaller than the potential surplus, strictly defined as the difference between the opportunity value of gas and the opportunity costs of the factors of production applied (where opportunity costs are determined by the perfectly competitive and riskless markets of economic theory, in which agents wield no power but have full and accurate information about all relevant aspects of the market together with perfect foresight). The gas industry is marred by uncertainties. Information is incomplete and inaccurate. Competition between companies is far from perfect, and every oil company, even if it is small, wields a certain amount of market power. There are thus both major risks and significant imperfections. For these and other reasons oil companies engaged in gas development upstream naturally expect a return on

---

[4] Normal profit in the Marshallian sense, i.e. the reward of capital in a perfectly competitive economy.

their capital markedly higher than mere competitive opportunity costs. They may expect a real rate of return of at least 20 per cent as against the 8–10 per cent nominal rate which prevails in today's money markets. Similarly commercial banks, where willing to finance part of the transmission infrastructure, are likely to offer loans at higher interest rates than the base opportunity rate in order to cover perceived risks. And where private firms are involved in gas development downstream they would also expect higher rates of return than the perfectly competitive and riskless opportunity costs of money.

Assume a gas project in a Third World country where upstream development and production is undertaken by an international oil company, and where transmission and downstream activities are in the hands of parastatal enterprises. Part of the investment in pipelines and in the distribution/marketing facilities downstream is financed by foreign loans. In such an instance, the true economic surplus (per unit) accruing to a country is:

> total opportunity value of gas produced
> plus government fiscal take from foreign producers
> minus sums paid to foreign producers
> minus interest
> minus total opportunity cost of other factors of production.

Usually this economic surplus will not correspond to the cash surplus generated by the project for the government. There are two main causes for the discrepancy. First, the gas price P_c charged to consumers may be different from P_r, and secondly the prices of other factors of production or inputs are unlikely to be equal to their true opportunity costs. The cash surplus is equal to:

> value of gas sales to end-users
> plus government fiscal take from producers
> minus sums paid to foreign producers
> minus interest
> minus actual payments for other factors of production.

There are three cases to consider. First, where the economic surplus is equal to the cash surplus. In this case the actual surplus accrues entirely to the government in cash form. It is part of government revenues and may be redistributed to the economy through budget expenditures. Secondly, where the economic surplus is smaller than the cash surplus. This arises when the gas price to consumers is higher than P_r and/or when factors of production are paid less than their opportunity costs. The difference is akin to an additional tax levied by the state on consumers and/or inputs. This is a rare case in developing countries; but it is important to see clearly, should such a case occur, that the cash surplus would then overstate the benefits accruing to the economy from the gas project. Thirdly, where the cash surplus is smaller than the economic surplus. In this case some of the economic surplus is appropriated by consumers and/or suppliers of inputs before reaching the government. The appropriation, in effect, is through implicit subsidies. This is perhaps a more common instance. It is important to note than in this case the benefits of the gas project accruing to the country are understated by the cash surplus obtained by the government.

Let us now turn to an important component of the cash surplus of a gas project which accrues directly to the government, namely the fiscal take from the producer's income. At issue here is the fiscal regime for gas upstream. It is important always to distinguish two cases. In case (a) the price paid to the producer for gas is lower than the opportunity price of gas to the economy. In such an instance there is an *implicit* tax levied on the producer equal (per unit) to the difference between P_r and P_p. More generally, when the price regime is not market related prices become fiscal instruments, used to claw back part of the producer's surplus (or, in rare cases, to provide the producer with a subsidy). We have argued that such pricing regimes are undesirable because they interfere with the allocative function of prices. To recap. The two economic objectives of ensuring an efficient allocation of resources and an equitable distribution of income are distinct from each other. Each of these objectives is best served by the use of distinct policy instruments: the pricing system is the preferred instrument for allocative purposes, and the fiscal regime (both taxes and subsidies) for distributional objectives. If prices are used to raise taxes or to subsidize economic agents they inevitably become distorted, and cease to reflect the opportunity costs of resources used up in production, or of goods and services marketed for consumption in the economy. We are fully aware that governments face great difficulties in correcting a price system which has already been distorted by past policies, and understand their predicament when faced with this problem. Precisely because of these difficulties, it is important to urge them *not* to distort prices from the outset when considering a new project or a new arrangement with producers or other economic agents. This is why we emphatically recommend case (b), where the price paid to gas producers is the opportunity price to the economy. In such an instance there is no implicit tax or subsidy. The producer and the government jointly share the price risks: they will both obtain a share of the upside potential should it materialize, and they will both share the burden of a price decline. When P_p is equal to P_r, the sharing between company and government of the surplus that emerges upstream is exclusively and explictly operated by the fiscal regime.

The fiscal regime for upstream gas can take different forms. Two conventional forms are of particular interest because they are adopted by a majority of countries: production sharing (e.g. Egypt, Tanzania) and OPEC concession-type regimes where the government fiscal take is not through an output share but rather through royalties and tax on a specified concept of taxable income (e.g. Nigeria, Tunisia). There are other systems, particularly joint-venture arrangements, where part of the fiscal take is translated into 'carried interest' provisions.

Production Sharing. In this arrangement gas output is typically split between the producing company and the government in three parts: (a) a portion of the output accrues to the company in compensation for costs incurred (cost recovery); (b) a proportion of the remainder accrues to the company as profit element; and (c) the rent is appropriated by the government and may be construed as the government's share of the gas rent.

OPEC Concession-type. Here, the government takes a royalty (usually in the range of 8–20 per cent of the value of the output) which is often treated as an expense for other fiscal purposes. The government also charges income tax on some concept of taxable revenue, usually the gross value of production minus royalties minus allowable expenses. The income tax rate may be as low as 40–45 per cent or as high as 70–85 per cent.

These are the bare bones of the two fiscal arrangements. Actual stipulations are infinitely more complex, but little would be gained in this context from an analysis of all the possible variations and details. Rather, our analysis will concentrate on the incentives/disincentives which these conventional fiscal regimes may entail. At this point, it is useful to raise the fundamental question of purpose: what are the objectives of the tax system? The simple answer is that the main aim of a fiscal regime is the maximization of government revenues, given the incentive constraint. The rules determining the fiscal take should therefore be such as to persuade the producer that: (a) in normal circumstances expected profits will be sufficient to justify involvement in the gas project; and (b) a share of the 'upside potential' of the project will be obtained. Producers feel that the latter is an important inducement, given that they bear a share of the downside risk. The fiscal regime should also ensure that the development of small fields, which may generate an economic surplus to the country, is not discriminated against through the imposition of a heavy fiscal burden on producers. This discrimination inevitably occurs when the fiscal regime takes, implicitly or explicitly, the form of a proportional tax on all fields irrespective of their characteristics and size. A proportional tax is inherently regressive, and one of the main difficulties in designing a gas fiscal system is to make it progressive.

In order to assess the impact of actual gas fiscal regimes on development prospects, we have undertaken a simulation exercise which produced a number of interesting results. The exercise considered four onshore and four offshore fields with the following production capacities: 1, 2, 3 and 6 million cubic metres per day (35, 71, 106 and 212 mcf/d). The useful production life of all fields was assumed to be fifteen years with a declining production pattern after the fifth year. For the purpose of this exercise the oil price was set at $15 per barrel and the gas reference price at 70 per cent of the crude-equivalent price. The load factor was set at 75 per cent of peak capacity. Both price and cost inflation were assumed to be 4 per cent per annum during the life of the project.

For the offshore fields it was assumed that drilling and production are undertaken through the use of fixed platforms, that the water depth is approximately 100 metres (328 feet), and that gas is transported to the onshore treatment plant by a 40-km (25-mile) pipeline.

The total capital expenditure of the eight typical fields was calculated through a detailed theoretical cost analysis of wells, platforms, sea lines, treatment plants, etc. The results are shown in Table 3.1.

We also considered four alternative fiscal regimes, which bear close resemblance to the actual arrangements in four of our case-study countries. In two cases the regime is of the production-sharing type (one conventional regime and one designed to give additional incentives to producers) and in the other two cases the concession-type, with royalties and tax on a specified concept of taxable income (again, one conventional and one incentive-enhancing regime).

Table 3.1: Estimated Capital Expenditure on Fields in Simulation Exercise. Million Dollars.

Case	Capacity	Capital Expenditure	
		Onshore	*Offshore*
A	1 mm³/day	28.4	89.2
B	2 mm³/day	40.9	110.4
C	3 mm³/day	50.6	124.2
D	6 mm³/day	79.0	160.7

These data were used to compute (for the four onshore and the four offshore fields, and in each case for the four types of fiscal regime) the internal rates of return and the payback period of the project before and after the fiscal take. The net present value in real terms of the project's gross revenue, discounted at medium cost of money (10 per cent), costs and government take, and the government share of the net income of the project over its lifetime (all in present value terms) were also calculated for the thirty-two cases. Sensitivity analyses were then carried out. The summary results are presented in Table 3.2.

The most noteworthy of these results are as follows. First there is a striking similarity between the impact of production-sharing and concession-type regimes on the after-tax performance of all the fields considered. This may be surprising at first sight, but closer examination suggests that production-sharing and royalties/income tax arrangements (resembling the OPEC historical systems) have in fact a similar underlying fiscal structure as far as the gas business is concerned. Secondly, we find that the after-tax performance of the projects examined is not very sensitive to the type of production-sharing agreement applied, be it the conventional or the incentive-enhancing type. Similarly, there are no striking differences in after-tax performance when the two types of concession agreements are compared. In other words improved fiscal arrangements of either type, although meant to give the investor greater incentives to develop a gas field, do not seem to improve sufficiently the after-tax performance of the 'unattractive' projects to justify their development. Thirdly, it can be seen that all the fiscal regimes considered in this exercise discriminate against small onshore fields and all but the largest offshore fields. There are cases where the internal rate of return and other performance indicators (before tax) do not justify the project. But in other instances the before-tax performance indicator would have made the project feasible if it were not for a fiscal burden. In other words none of the fiscal regimes considered appears to distinguish between small and large, or between onshore and offshore fields. It seems that the fiscal take rises in a linear manner across fields of different size and of different cost structure. This points to a need for improving the fiscal system in order to ensure the development of economically profitable fields which are at present discriminated against. This seems to support the argument in favour of progressive fiscal regimes. In many countries the gas taxation system is regressive, and this feature inhibits development – particularly offshore, where a large proportion of incremental reserves are likely to exist.

Table 3.2: Results of Simulation Exercise for Different Fiscal Regimes.

Case	Internal Rate of Return (%)				Payback Period (Years)				Government Take as Percentage of Net Income in Present Value Terms	
	After Tax		Before Tax		After Tax		Before Tax			
	Onshore	Offshore	Onshore	Offshore	Onshore	Offshore	Onshore	Offshore	Onshore	Offshore
(a) Conventional Production-Sharing										
A	16.2	<0.0	46.3	3.3	6.9	>19.0	4.6	11.6	88.0	>100
B	22.1	<0.0	60.7	15.7	6.1	13.9	4.1	7.6	83.0	>100
C	25.6	6.4	68.5	23.1	5.7	9.5	3.9	6.5	82.0	>100
D	31.2	13.6	81.5	35.8	5.2	7.7	3.7	5.5	80.0	91
(b) Conventional Concession-Type										
A	20.6	<0.0	46.3	3.3	5.4	18.3	4.6	11.6	83.0	>100
B	26.9	5.9	60.7	15.7	4.9	8.8	4.1	7.6	81.3	>100
C	30.3	8.8	68.5	23.1	4.6	7.8	3.9	6.5	80.7	>100
D	35.6	15.9	81.5	35.8	4.3	6.6	3.7	5.5	80.0	87
(c) Enhanced Production-Sharing										
A	24.6	<0.0	46.3	3.3	5.6	13.9	4.6	11.6	73.0	>100
B	29.8	8.5	60.7	15.7	5.2	8.2	4.1	7.6	74.0	>100
C	32.7	9.2	68.5	23.1	5.0	8.2	3.9	6.5	75.0	>100
D	37.7	20.4	81.5	35.8	4.7	6.4	3.7	5.5	75.5	74
(d) Enhanced Concession-Type										
A	25.3	<0.0	46.3	3.3	5.4	18.6	4.6	11.6	71.0	>100
B	33.3	7.1	60.7	15.7	4.9	8.4	4.1	7.6	68.0	>100
C	37.4	12.3	68.5	23.1	4.6	7.3	3.9	6.5	67.5	88
D	43.7	19.8	81.5	35.8	4.4	6.4	3.7	5.5	67.5	75

Our recommendations for designing gas fiscal regimes are:

(a) Avoid unnecessary complication. The tax system should be as simple and as transparent as possible. Usually neither company nor government welcomes such a recommendation because each side believes that complex or obscure stipulations can work to its advantage. (The rejection of the Part Report by *both* the UK government and North Sea operators, ultimately because the proposed tax system for UK oil production was both straight-forward and transparent is an interesting illustration.) We believe that the benefits to both parties of tax systems devoid of byzantine complications exceed the specific advantages to any party of unnecessarily complex regimes. This applies especially to their long-term relationship, in which there is less room for argument over interpretation, less pressure for constant revision, less litigation and so on.

(b) Avoid the multiplication of taxes. There are no good reasons to charge signature bonuses, licence fees, rent or bonuses on concession acreages. However, companies should expect higher rates of royalties or income taxes when these additional taxes are not levied than when they are.

(c) Provide for accelerated recovery of capital expenditure in the first years of production. This does not mean that the cost/recovery rate should always be 100 per cent (as against the typical 40–50 per cent of current regimes), but that consideration should be given to reducing the payback period to a maximum of three years, depending on the characteristics of each field.

(d) Adopt the principle of progressive taxation. This could provide incentives for the development of smaller fields which are only marginal under a proportional tax regime. Net income would be shared by governments and companies on a sliding scale. Both the volume and the price risks are thus shared. The government's share will be low when the net income of the project is small but will rise, and could indeed be made to rise very sharply, where income increases. A well-designed progressive tax system may preclude the need for the much-hated windfall tax on exceptional profits. Companies, however, should accept that in certain circumstances the case for such a tax does arise. A windfall is 'manna falling from heaven' and is for everybody to share; since the government is the lion in the land it gets the lion's share.

(e) Ensure that the fiscal system is as neutral as possible with regard to allocative decisions. This is achieved to a great extent when producer prices are market related. However other interferences with the efficiency of resource allocation may occur where the tax system encourages 'gold-plating' in exploration and development, or involves many loopholes, leading companies to modify their normal economic behaviour in order to take advantage of tax breaks. In this context we do not share the opposition of certain international agencies and of some experts to the concept of the ring fence. It is perfectly logical to tax gas production on a field rather than on a company basis. The reason is that exploration and development decisions in any given area should be made on own merits and not be influenced by tax allowances against profits made in another area or field. The profitability of one field is no indicator of the prospects of another and should not be allowed to play a role. It is important however to stress that if a fiscal ring fence around a field is used, this field-by-field approach should also be extended to all other aspects of the contract and taxes. It should also allow for much greater flexibility in the tax regime, to suit the diverse and sometimes changing conditions of each field.

It is not beyond the ability of fiscal experts to design a regime on the basis of these principles. Our own preference is for a system which combines the best features of the production-sharing and of the resource-rent tax approaches.[5] The advantages of production-sharing lie in its provisions for cost recovery and for tax payment through free gas deliveries to the government. This is practical, since the government in most developing countries purchases the whole output of gas destined to the internal market. The advantages of the resource-rent tax are that it protects the smaller fields, guarantees a basic rate of return, and can provide for the progressive sharing of excess income between government and company. It is unfortunate that the resource-rent tax has acquired a bad reputation with companies through the zeal of some of its promoters, who presented it as a means for skimming the entire rent, leaving companies with a predetermined rate of return. This feature is not however an inherent feature of such a tax since the rate of tax can be set at any level, not necessarily at a punitive 100 per cent.

---

[5] A resource-rent tax is based on the *ex post* profitability of a project, and escalates with increasing discounted cash flow returns on investment. This approach minimizes front-end risk for the investor and accelerates cash flow, but still allows the government to obtain a high share of the rent from gas production.

4 PLANNING AND FINANCING A GAS PROJECT

4.1 Introduction

Previous chapters have described the various problems which often stand in the way of developing gas fields in the Third World for internal markets. In Chapter 1 we assessed gas in relation to alternative fuels, and described the features of the main potential markets for gas in the internal economy. The strong conclusion is that despite its premium qualities gas can only penetrate the energy market in sufficient quantitites to justify its development costs if it is largely applied to non-premium uses (usually the power sector). Gas will thus fetch prices that are more closely related to the heavy end of the oil product barrel than to those of the 'noble' petroleum products, in other words it will fetch prices lower per btu than crude oil. This is one of the many reasons why governments' attitudes towards gas, and expectations about its benefits, should not be modelled on their oil policies. Gas is different to oil in many respects, and approaches to its development need to take full account of its specific features.

In Chapter 2 we portrayed two main protagonists of gas development: the international oil company and the government. Both share, albeit for different reasons, an interest in those gas projects capable of yielding profits to the company and economic benefits to the country. However these two potential partners have different aims, some divergent interests, different approaches to decision-making and different administrative structures. They both face financial and manpower constraints, but of different sorts. Consequently there is scope for misunderstanding and even conflict between the two sides in their most basic assumptions and approaches to a project.

In Chapter 3 we examined the three difficult problems of volume, price and the sharing of the gas rent. There are uncertainties about volume on both the supply side (reserves, discoveries, production difficulties) and the demand side (size and growth of potential markets). These difficulties are significant, because gas exploitation is tied to specific outlets; hence the need to find a secure market before extensive exploration is carried out, and the concomitant difficulties of planning markets before the extent of reserves is known. The pricing issue is both conceptual and political. Although economic analysis enables us to define appropriate concepts for pricing gas to producers and consumers and for establishing its opportunity value to the economy, governments often find it difficult, for political reasons, to base their pricing policies on these concepts. Finally the appropriation of a share of the economic surplus by the government in ways which provide sufficient incentives for the participation of companies and other economic agents in the project raises the complex issue of taxation. Governments owe it to their countries not to concede to companies more than is necessary to induce them to invest; but for governments to demand too large a share of the economic surplus may inhibit the development, depriving the country of all benefits. Similarly companies that are too greedy in their demand for a share of the pie might well fail to obtain a contract and thus be deprived altogether of the opportunity of making a profit. Sharing the rent calls for the exercise of good judgement, technical competence in gas fiscal matters and a realistic approach to bargaining by all the parties involved. Further problems mentioned in various parts of this book include the issues of very low domestic energy prices, and of convertibility of earnings from gas sales within the country to hard currency.

While this array of problems and difficulties may seem daunting, none of them is in itself insuperable. The aim of this chapter is to outline a course of action for successfully avoiding the various pitfalls. The solution proposed involves the following. First, a number of policy and legislative measures to be taken by governments from the outset in order to establish simple, clear, yet flexible terms of reference for gas developments (Section 4.2). Secondly, the *joint* planning of the gas project by all the parties involved – government and producing company – in close consultation with the financial institutions invited to provide loans (Section 4.3). If a gas project is approached and planned in an unstructured manner then the chances are very high that it will collapse at an early stage, as one problem or another is encountered.

Of course, the joint planning approach entails certain risks. Some companies and some governments might so reveal themselves in the planning process that working together might not be beneficial. Differences in working styles can also create difficulties. We believe, however, that the existence of such risks is not a valid argument against joint planning as such. The problems that may arise between the parties during joint planning of a project would in any case emerge at the negotiating stage. Surely it is better to face them when working together on the many concrete problems involved in planning a project, rather than confronting them around the negotiating table.

4.2 Initial Measures: Model Gas Clause

It is very rare for companies to explore in the Third World specifically for gas, because of the problems that have traditionally been associated with its exploitation. Companies hope to find oil, for which there is a ready world market. The consequence of this attitude has been that many exploration licences have contained no provisions for gas, beyond the statement that both sides will discuss what to do should gas be found. This lack of a framework for gas will always lead to delay in developing a gas find, since so many fundamental questions remain to be discussed between company and government; in gas-prone areas it may also actively discourage companies from exploring for oil, since they will anticipate such delays, and are at the same time aware that most governments are no longer prepared to look on gas finds as 'dry holes'. It is therefore well worth while for the government to draw up in advance the ground rules according to which gas finds will be treated. This may be either in the form of legislation, or a model contract which can serve as the basis for detailed negotiation with interested companies.

The purpose of such legislation or gas clause in an exploration contract is not to lay down precise terms. These must always depend on the size of the field, its location, the quality of the gas, its distance from potential markets, the exact timing of the discovery in relation to the country's economic development, and so on. The level of detail that is appropriate will vary according to the stage of gas development in the country; if no gas has ever been used (as for example in Tanzania), there is no existing infrastructure, no precedent for pricing, and no body of experience in gas management it is very difficult to give more than very broad statements of principle; in a country such as Pakistan, on the other hand, there is scope for defining policy in much more detail in advance of gas discoveries, thanks to the existing markets and gas experience.

The gas clause should nevertheless define, at least in broad terms, the following:

(a) The pricing principle. Our recommendation is that gas prices to the producer should be determined according to market-related criteria; and this general principle can be stated, without the need for entering into detail over the exact netback formula for calculating the price.

(b) The broad principles (other than price) governing sales to the downstream agency – whether take-or-pay and delivery obligations will be involved, the method and currency of payment, and the limitations (if any) on convertibility. The precise specification of take-or-pay obligations should be determined through negotiation project by project, when all the relevant characteristics of both the gas field and the market outlet have been fully appraised.

(c) The gas clause must also define the main features of the fiscal regime, be it production sharing, joint venture with carried interest, concession agreements with royalties and income taxes, resource-rent tax, etc. Our recommendations on this issue were spelt out in Chapter 3. The extent of detailed specification depends on the stage gas development has reached in the country concerned.

(d) It is useful to provide in the gas clause, from the outset, for flexible exploration terms, and to indicate the government's willingness to extend terms or to allow companies to postpone development for agreed periods of time, depending on the findings of the joint planning exercise outlined below.

(e) A most important component of a gas clause must always be provision of a clear procedure for joint assessment of the find by both company and government, and a mechanism for ensuring that negotiations at least start in the right manner. The planning process which synchronizes appraisal both upstream and downstream, and the various steps that must be taken before the final declaration of commerciality is made and development is commenced, should thus be defined in advance; this will simultaneously give a higher chance of gas finds being exploited, and encourage companies to participate in exploration in the first place. Our suggested pattern for such planning is described below.

4.3 Planning a Project

A fundamental requirement of any project has been shown to be the need for careful planning, to synchronize the upstream and downstream aspects of a project, since these are so interdependent. This in itself leads to the need for very close co-operation between company and government, since they are normally responsible for the upstream and downstream respectively. Their interdependence is heightened by the need to recognize that unless they both benefit from the project, and unless each party sees that the other is not taking undue advantage of its position, then the project will not be undertaken and neither will benefit. The planning process is therefore more than a mechanical exercise. It needs to foster an unusual relationship between the agents involved from the very earliest stages of a project, and calls for a high degree of sensitivity in negotiations.

We believe that this is best achieved by the joint planning of the gas project. Over and above this general approach to projects there is a series of specific recommendations which need to be inserted at particular stages of the planning process. We hope to pinpoint when the various parts of the pricing policy need to be articulated; when market promotion becomes necessary; when financing should be arranged; and so on. Of course each individual project will vary in detail, but we believe there is sufficient commonality between gas projects to make such an exercise both useful and illuminating.

A number of distinct phases can be identified in any gas project. The breaks between phases are marked by the opportunity for taking a decision whether or not to continue with the scheme, whereas within each phase activity is more or less routine. The very first step to be taken is to initiate exploration for gas and/or oil. Once gas is discovered a preliminary decision needs to be taken over its commerciality; this is the phase which in our judgement is often mishandled, and needs a fresh approach. If the gas find is judged at this preliminary stage to be worth developing, then the company will need to carry out further work to firm up reserves, and the government will need to plan its marketing strategy in more detail. If both these activities bring good results then negotiation between company and government over the details of production and sales contracts needs to be pursued. It is worth spelling out these phases in detail, in order to demonstrate how all the earlier conclusions and recommendations in this book relate to a project, and are interrelated with each other.

(*a*) *Is a Project Worth While?* There is a general assumption that companies, governments and financing agencies will all take a decision that a project is commercially beneficial before they become involved in negotiation with the other parties. They will each decide whether they individually can gain enough from the project to justify the inevitable risks involved. Our argument is that such a process is often counter-productive. It cannot reach sensible conclusions, since no party is independent of the policies and decisions of the others. By building in preconceived ideas of how each party will receive its share of the benefits it can also make negotiation unnecessarily difficult. An adversarial attitude is often taken up at this very early stage which is very difficult to overcome later.

It is precisely in order to remedy this situation that we suggest that the initial assessment of projects be carried out as a joint undertaking. The priority is to establish whether there is the makings of a viable project, which is capable of generating large enough returns to allow a share-out between the gas producer and the government. If the project is so marginal that there is no hope of satisfying both sides then there is no point in pursuing negotiations over the exact shares of each. Using opportunity value as the basis of evaluation the scheme as a whole should be assessed, including upstream and downstream elements. Preliminary estimates of extraction costs, distribution costs, and size of markets can generate an estimate of overall profitability and benefits to the country, without apportioning shares of the profits between the agents involved. It is insufficient for the company to satisfy itself over reserves and the government over the market; each side must persuade the other of the viability of the project from its own point of view.

Such early studies are often carried out by consultants for one side or the other. The findings and recommendations of a consultant working for one side are often mistrusted by the other side however, not least because consultants are usually suspected of telling their clients what they want to hear. It is therefore important that the studies be carried out by someone who is equally respected by all those concerned. One way of achieving this would be for the studies to be jointly commissioned; alternatively there may be a useful role for agencies such as the World Bank in carrying out initial evaluations of projects. Contractors should not be entrusted with the evaluation, since they will merely introduce a further competing interest; they should simply be consulted on the issue of costs. A neutral facilitator may play a useful role to spearhead the work; but ultimately the aim is for the company and government to build trust in each other and to learn to understand the problems and behaviour of each other.

If the conclusion of these early studies is that there is no viable market for gas at that time the judgement should be accepted, but not taken as a final dismissal of all hope. The situation may well change over time, as markets develop in new ways and the price of alternative fuels alters. The possibility of a project should be kept under regular review. If however the initial assessment is favourable, then the next stage should be broached.

(*b*) *Appraisal Stage.* At this point both company and government should move on to more detailed analyses. The company will become involved in detailed field appraisal, while the government will pursue market feasibility studies, with the help perhaps of consultants from major gas companies. Assuming that earlier hopes are confirmed by these studies, then the government will need to start drawing up sales agreements with gas customers, and also seeking finance for infrastructure. It will also be important at this stage that the government takes care to bring together the various interests and pressure groups within its ranks, in preparation for more detailed negotiation. As discussed in earlier chapters, the oil ministry, the power authority, the Ministry of Finance, the national oil company and the planning authority will represent a range of opinions and of enthusiasm for gas. It is very easy in this situation for one arm of the government to work against another, whether consciously or unconsciously. It is also essential for the government to have a unified voice in negotiations with oil companies. This may be facilitated by the use of an official Gas Mediator as discussed in Chapter 2.

Above all these studies should not be seen as self-contained exercises, of interest only to the party most directly involved. Further details of gas reserves may have a profound impact on market planning; and market opportunities always remain extremely important to the company. There is thus a constant need for communication and sharing of information. The decision to declare a discovery commercial, and to proceed with development must be a joint one.

(*c*) *Commercial Development.* Once a firm decision has been taken to develop a field, then details of the sales contract between oil company and government must be negotiated. It is at this stage that the share-out of benefits is determined. The general principles should have been outlined in the exploration agreement, but details will still need to be agreed upon. For example there should be an acknowledgement that prices will be market related; but the exact formula will need to be calculated, as also the fiscal regime which accompanies it. While there will of course always be room for dispute in such discussions the chance of a favourable outcome will be much enhanced if both sides have already been working successfully together in the earlier stages of joint planning, as outlined above, and perceive that a solution will be to their mutual advantage.

4.4 Financing

External capital flows for gas projects can include:

(a) Funds from international oil companies, either as equity contributions or as loans to a local subsidiary;
(b) Export credits obtained by the contractor;
(c) Commercial bank loans to a local operating company;
(d) Loans from multilateral development banks;
(e) Official development assistance either from multilateral aid agencies or from governments.

The source relevant to any particular project will differ, depending on whether an international or a national oil company is involved upstream, on whether the country is eligible for official development assistance and loans from development banks, and on the nature of enterprises involved with transmission, downstream distribution and marketing.

Consider the typical case where an international oil company operates upstream and parastatal enterprises are in charge of transmission and the internal market. The financing of exploration and development is then the responsibility of the international company. Historically financing of what they perceived as 'good' projects has not been a serious constraint on investment by major oil companies. But financial resources are not unlimited: companies operate within investment budgets, whose sizes depend both on cash flow assessments and on the range of investment projects to be undertaken in the budget period. Each potential project competes with others for these investment funds; and gas in developing countries, particularly gas for the internal market, is at an initial disadvantage in this competition. Today, as discussed in Chapter 2, the cash flow situation of several US oil companies is strained, partly because of lower oil prices and partly because of mergers and restructuring. These companies have slashed their investment budgets, and certain gas projects which they may have considered in happier days will be flatly turned down today. However, other oil companies are in a good financial situation, with low debt/equity ratios and/or large cash reserves. These companies are likely to take advantage of the decline in drilling and other investment costs, which have paralleled the oil price fall; the prospects for gas projects through these companies have not deteriorated. In fact, we noted in interviews that they have a definite interest in gas development in the Third World.

Very few developing countries can expect to obtain external finance from private banks, under today's conditions, for the gas transmission system or for downstream facilities. Direct loans to a national oil company for these purposes, when the gas project is for the internal market, are extremely unlikely, although there will be occasional exceptions. For example, Chase Manhattan has recently made a loan to the Argentine Central Bank which is earmarked for private sector gas development. This particular project should be very profitable and Chase has had a long involvement with the company concerned. A country can always attempt to raise a multi-purpose syndicated loan, rather than seek a credit for a specific project that is unattractive to the banks. Syndicated loans were very fashionable between 1976 and 1983; indeed a large part of Third World international debt was created by such loans. For obvious reasons, banks are becoming much more cautious today, and most new loans to the Third World are to restructure their old debt and to enable them to service it, rather than to undertake new projects. We should qualify all this however by noting that European and Japanese banks are generally more positive than US banks on lending to Third World countries (other than in Latin America).

This suggests that gas projects for internal markets in developing countries depend to a significant extent on the willingness of multi-lateral development banks and official aid donors to provide the sums required (above what the government can manage to raise domesti-cally, through foreign exchange allocation in the national budget, or from limited loans). Finance is required for transmission and for the downstream market. International oil companies have some interest in becoming shareholders in downstream investment, but are not generally willing to provide more than a small proportion of the finance required. The World Bank does finance energy projects in the Third World, and since 1976 has consistently expressed an interest in gas. The Bank has mainly financed pipelines, technical assistance and the drilling of appraisal wells. In the fiscal year 1985 it committed '$525 million for gas operations. It is currently funding 16 natural gas and 2 liquefied petroleum gas projects in 12 developing countries, for a total Bank commitment of $1.7 billion. New lending for gas projects in fiscal years 1986, 1987, and 1988 is tentatively projected at $1.8 billion for 26 projects and technical assistance packages in 24 develop-ing countries.' (*Finance & Development*, June 1986, p. 12.) There is clearly a commitment and funds are indeed made available; but considering that most other sources are almost dry, that the number of developing countries with gas prospects is large, and that gas development is very capital intensive, we feel that the sums commit-ted by the Bank are not commensurate with the importance it itself gives to gas development.

Although we have listed sources of funds separately, generally a gas project will be financed by a combination of different sources. The oil company will seek export credits as well as providing internal funds. Some companies may even seek bank loans, but are unlikely to obtain them for gas projects under current circumstances. The involvement of the World Bank, other development banks and aid donors has greater significance for the project than simply the sums involved in the loans. Their presence can encourage others to participate in the financing, because of their influence with the host government (should conflict arise) and because their participation may indicate favourable assessments of both the project and the country's economy.

A project's financial package may be put together more easily if some insurance is provided. Some governments have their own insurance schemes for investors/exporters. The Overseas Private Investment Corporation (OPIC) in the USA was established to encourage and support US private investment in friendly countries. In recent years OPIC has become more active in insuring petroleum projects (now 25 per cent of its risk insurance portfolio). OPIC's insurance coverage has been biased towards the more developed countries, but this simply reveals the pattern of overseas oil investment by US companies. However, OPIC is already heavily exposed in Latin America; it is opposed to financing projects in which the public sector plays a major role; and does not look at gas projects very favourably at present. At least, so we gathered in interviews.

There are multilateral insurance schemes such as MIGA. This proposed institution has not yet begun operations, but may prove promising. It will insure for political risks in developing countries, with coverage for:

(a) currency transfer (but not exchange rate depreciation);
(b) expropriation;
(c) breach of contract;
(d) war and civil disturbance;
(e) other agreed non-commercial risks.

MIGA may help to stimulate private direct investment in developing countries, but the sums initially involved ($1.5 billion) are small, considering the magnitude of gas investments and the fact that gas is but one activity among many eligible for insurance.

There is clearly an urgent need for significant innovation with regard to the financing of gas projects. The IFC is taking some small steps in this direction, through equity investments in private oil and gas companies in the Third World, and through a new Guaranteed Recovery of Investment Principal (GRIP) programme, where the IFC acts as an intermediary for equity investments. Under GRIP, the IFC effectively guarantees the capital value of the equity, but shares capital gains and dividends with the investor. Another route, seen by some observers as a potentially effective mechanism for the promotion of gas development, is the sectoral adjustment lending programmes developed by the World Bank and regional development banks. This programme is an effective mobilizer of capital through co-financing; it involves the Bank as an active participant seeking to harmonize the conflicting interests of the government and the private sectors; it requires the country to define economic (and by inclusion, energy) policies clearly and coherently, which can only be of help to gas development; finally it draws the Bank into the planning of the gas project as a long-term package of interrelated investments with links to several sectors of the economy. All these features of sectoral adjustment lending are extremely positive, because they relate financing to sectoral development, economic policy, institution building and to the co-operation of public and private enterprise. There are serious difficulties with it, however, arising from the most important, yet most controversial aspect of sectoral adjustment lending: the close monitoring of policy which can impinge on sovereignty, thus causing frictions which result in the interruption of disbursements and uncompleted projects.

We are inclined to think that only a revival of traditional project aid, the original objective of development banks, backed by sound economic appraisal of projects, can solve the financing problem of gas development. This precludes neither the participation of the Bank (and sister organizations) in promoting and planning gas projects, nor its involvement in policy reform. As argued earlier these roles are also crucial for the success of gas development.

5 CONCLUSIONS

The purpose of this study has been to look at the scope for using natural gas in internal markets of developing countries. We have excluded LNG, and also the special case of countries with very large oil reserves, where special economic considerations apply.

It is not automatically to a country's advantage to develop its natural gas resources simply because they exist. The first step is therefore to define under what conditions the country should seriously consider using its natural gas. The criterion to be used in this evaluation is the net opportunity value of domestic gas for the economy. This involves a comparison of the border price of the substitute fuel (whether this be the fuel that would otherwise be imported, or the fuel that will be released for export) with: (a) the opportunity cost of the domestic resources used up in developing and extracting gas, (b) the sums to be paid to foreign producing companies, foreign contractors and banks for their services, and (c) the opportunity cost of non-optimal depletion. Perceptions about future energy prices will obviously have a strong bearing on the conclusion of such studies. Thus, under current conditions of uncertainty regarding the long-term movement of oil prices, fewer countries will reach a positive conclusion than would have been the case when oil prices were assumed to rise. It is important to stress once again that the country's initial interest in developing a gas reserve should depend almost entirely on an appraisal of the benefits it can bring to the country's economy. This preliminary step should always be taken before any further considerations are raised.

As discussed in Chapter 4 this first decision as to whether to pursue a gas discovery should be taken in close conjunction with the oil company which made the discovery. This may be either a national company or a foreign one, with the latter being more common. The decision should also involve at an early stage the expert help of gas utility companies, which can provide advice on a consultancy basis, and perhaps other inputs, in identifying, appraising and developing downstream markets.

While the government and the oil company may be the principal players they are not the only parties involved. In most developing countries the only way of financing gas pipeline construction will be through aid programmes. Agencies such as the World Bank have displayed interest in encouraging the use of gas, but there is a lot more scope for positive action on their part. They should play a significant role in providing finance for the pipeline systems of good gas projects, recalling that aid for big infrastructural projects is their traditional function. There are a number of additional ways in which their participation can be very helpful: they are in a good position to assist the country in the very early appraisal of whether or not a gas reserve should be developed, and they may play an important role in identifying and promoting potential markets for gas. In short, there is a strong case for an early involvement of international aid organizations in the conception and appraisal of the project.

Where development upstream is undertaken by a foreign oil company, there is a dichotomy between two groups of agents with very different interests and objectives. Whereas the country is concerned to bring the greatest benefit to its economy, in a context set by political and strategic interests, the company is primarily interested in generating profits. While there is a divergence of specific interests and expected rewards however, it is important to remember that in the final analysis both parties are working towards the same end – a successful gas project. This is not a zero-sum game. It is fruitless to try to subordinate the interests of one group to those of the other, by arguing that one set of interests is more important or more soundly based than the other. If this line is followed then the chances are that nothing at all will be achieved, and neither party will gain anything. Rather, the aim should be to try to recognize the shared interest of both groups in the success of the project. If both company and government are prepared to recognize each other's different objectives and constraints then there is a possibility of finding common ground – always remembering that if a successful project can be brought about, then both parties stand to benefit.

A further prerequisite to a successful project is the recognition by all the parties involved of the peculiarities of gas as a resource. These can be characterized under three headings: the long-term nature of gas projects; the interdependence of all the factors involved in a gas project; and the requirement for a high level of trust between company and government which arises from these first two. These intrinsic features of gas development bring particular pressures to bear on the negotiating process. They also reinforce the need for co-operation and communication between the government and company, which was alluded to above.

First, it has to be recognized that there is a major difference between the long-term nature of gas projects and most oil projects, leading to a need for long-term planning and for the adoption of a forward-looking strategic approach by all those involved. In the case of oil the world market can be assumed to absorb production and the downstream therefore rarely calls for major long-term decisions. There is no such cushion for gas; demand has to be carefully planned over the same time-horizon as the life of the gas field. Gas is also an unfamiliar fuel and time must be allowed for the conversion of both minds and fuel-using plants, thus for a slow build-up of demand which can eventually justify the initial investment. Hence there is a need for the government to plan and gradually promote the use of gas for a long time ahead, whilst retaining sufficient flexibility to absorb the inevitable unforeseen factors of a country's energy picture. Some form of long-term energy strategy is therefore necessary. Such a strategy would normally feature diverse sources and types of energy to ensure adequate security, with gas being just one amongst several energy sources. Put differently, the decision to develop gas resources involves thinking about the diversification of energy supplies as a long-term insurance against possible 'shocks'. The oil company, on the other hand, may see this dependence on specific long-term demand, which puts it at the mercy of future government actions, as an additional risk when compared to oil projects. If it assesses individual gas projects in competition with oil projects, the former may therefore seem of little interest. A gas project could be of far greater interest however if companies considered it in the context of a strategy of upstream portfolio diversification, providing the company with flexibility to meet unforeseen future events. Indeed there is evidence that those oil companies which maintain that they have such a strategy are in fact more active in attempting to develop gas projects in Third World countries. Both governments and companies thus need to look at gas projects within the context of some overall strategic aim, an unfamiliar stance to many.

Secondly, gas projects must be conceived as total packages. The upstream phase cannot be planned in isolation from the identification and promotion of markets, and pipeline financing also needs to be considered at a very early stage. There is thus a great need for synchronization of all the different phases of a project, as described in earlier chapters, with company and government working in close harmony. At the downstream level a complicated group of interdependent factors also needs to be taken into account. From the macroeconomic point of view the existence of markets for gas will depend on the overall economic development strategy of the country –

particularly the growth of industry, but also expected transport requirements and residential energy demand. Most importantly, a gas
project cannot be planned without giving detailed consideration to
planning for the power sector, since this is highly likely to provide the
initial bulk market for gas. The substitution of gas for other fuels will
of course also affect overall energy policy, as the pattern of demand for
oil products will alter. Once again this points to the need for a general
energy strategy to be drawn up, (or modified if one already exists). To
add to this multiplicity of tasks, it is likely that different groups within
the government structure will have differing levels of enthusiasm for
gas, and may even try to block its development. The appointment of
an agency or individual responsible for co-ordinating the government's position may prove helpful in this respect, as discussed in
Chapter 2. Rather than a sequence of discrete decisions therefore, a
gas project requires an integrated decision-making process. Upstream
and downstream (company and government) are totally interdependent at all stages, and the government itself will need to carry out
extensive co-ordination within its own ranks.

The result of these two factors – that gas requires long-term planning and exceptional co-ordination of its phases – is that company
and government are indeed forced into a uniquely close relationship.
They need to work in very close contact, over an unusually long
period of time. The need for the two sides to be clear over their
mutuality of interests is thus re-emphasized from a different angle. In
particular it should be noted that gas projects are totally different
from oil production in this respect; unless companies are fully aware
of the need for a different approach then there can be little hope of
success.

For company and government to operate happily together will
require each to make concessions to the other, in order that the
interests of both can be served. Neither party can expect the other to
accept an outcome which maximizes the benefit to one party, leaving
the other with little reward. Some of the major concessions required
can be listed.

Governments are more likely to win the co-operation of companies
if they accept the following:

(a) The principle of market-related pricing. The gas price paid to
    producers should be calculated in relation to the substitute fuel,
    netted back to the well-head.
(b) That companies carry a large part of the risk in a gas project, and
    should therefore be allowed some of the upside potential of the
    project.

(c) The rent on the resource should be fairly distributed between the company and government, preferably through the taxation system.

Governments have made great efforts in the past few years to provide companies with these necessary incentives.

Companies on the other hand must bear in mind the following:

(a) There are political limits to the share of the rent that governments can offer them.

(b) More generally, governments have a number of objectives, not all of which are economic. Governments need to satisfy multiple interests. It is important that companies acquire expertise in political economy to enable them to appreciate these factors, and negotiate accordingly.

(c) There is an asymmetry of information, in the sense that companies are much more knowledgeable about gas. This is felt very acutely by governments, who fear that they may be outman-oeuvred by the companies due to their own lack of experience. Such feelings can cause considerable delays in negotiations; companies may be able to alleviate this problem by providing plenty of information, and possibly training. Above all their negotiators need to be sensitive to the problem.

(d) Companies must be, and be seen to be, concerned for the welfare of the country in which they operate. Attempts to dismiss the importance of side-effects of a project on the economy, when these may worry a government, are likely to be counter-productive. It always helps to give disinterested advice on broad policy issues relating to the project when called upon.

Overall it can fairly be said that both companies and governments experience difficulties with each other when looking at gas projects in developing countries. These problems are not insuperable however. The key to solving them is for each side to understand the reasons why the other side objects to its behaviour, since this may enable it to modify that behaviour in some of the ways suggested in this book. They also need to remember that ultimately they both share the same interest; namely the success of the project, which will allow both of them to reap a reward.

Finally one can ask, what are the realistic prospects for gas developments in the world today?

We have attempted to answer this question by undertaking two exercises. The *first exercise*, kindly conducted by members of the Energy Department of the World Bank on our behalf, attempts to evaluate the unit costs of gas development for a number of 'typical' onshore and offshore projects in the Third World. The model used cost data from actual projects undertaken in recent years in a variety of countries and financed by the World Bank. The following results are of interest. First, the costs of gas development upstream and downstream are as low as $0.70/mcf for a base case onshore field, assuming a 10 per cent discount rate. Setting the discount rate at 20 per cent raises the unit cost to only $0.83/mcf. For an offshore field, in the base case, unit costs are $1.51/mcf (10 per cent discount rate) and $1.86/mcf (20 per cent discount rate). The most unfavourable project in this exercise involves unit costs of $2.17/mcf in the base case. This exercise suggests that current fuel oil prices – say, $13 per barrel in 1987 – support most gas developments in the Third World, in so far as unit costs of development are concerned. The important conclusion is that costs are not the obstacle to gas development in the Third World. This confirms the view expressed in this book that the obstacles lie elsewhere. Secondly, unit costs rise considerably when developments are delayed. In fact this factor causes the greatest relative increase in costs (to illustrate, the cost estimates are more sensitive to 'delays' than to assumptions about the exploration success ratio). The message is therefore clear. Where the obstacles to gas development lie in the relationship between government and company, and delays in planning and negotiations occur as a result, new obstacles are likely to emerge in the form of escalating costs. Delays should therefore be avoided.

The *second exercise* evaluates the prospects for the twenty-nine countries listed in Chapter 1,[1] all of which are Third World countries with gas reserves and identifiable markets for that gas. Twelve criteria were applied in an attempt to gain a feel for the prospects of gas development. Some of the criteria used are value-laden and difficult to assess; others are more straightforward. These criteria were chosen because they include all the major factors identified in this book as common stumbling blocks to gas projects. They are: (a) gas reserves, (b) market potential for gas, (c) availabilities of commercial credit, (d) foreign exchange constraints, (e) willingness of international aid institutions to fund projects in the country, (f) existence of legislation or model clause for gas development, (g) the country's record on contract stability, (h) bureaucratic efficiency, (i) distortions in domestic energy prices, (j) existence of a national oil company, (k) current involvement of international oil companies in the country, (l) pro-

---

[1] Abu Dhabi, Afghanistan, Algeria, Argentina, Bangladesh, Bolivia, Brazil, Burma, Chile, China, Colombia, Ecuador, Egypt, India, Indonesia, Iran, Iraq, Malaysia, Mexico, Nigeria, Pakistan, Peru, Qatar, Saudi Arabia, Syria, Tanzania, Thailand, Tunisia, Venezuela.

motional drive of the government institutions likely to be involved in a gas project.

We found that whereas commercial financing is only rarely available, most countries do have access to aid funds, providing the agencies are prepared to lend for gas projects. Foreign exchange problems are also not as widespread as generally believed. On the other hand the distortion of domestic energy prices emerges as a predominant obstacle, rating poorly in nearly all countries. Bureaucracy is also a problem in a number of countries, often combined with opposition from the national oil company towards gas projects. There is therefore an urgent need to concentrate on finding solutions to these specific problems. But our assessment is that the difficulties are not insuperable. There are good prospects in many countries, particularly in China, Egypt, India, Indonesia, Malaysia, Pakistan and Thailand, and in a number of the Latin American countries which show potential for projects on perhaps a smaller scale.

It should also be borne in mind that this exercise of course looks only at those problems which lie within the remit of the government. This by no means implies that all the problems lie in that quarter. Rather, one important message of this book is that some of the obstacles to gas development lie in the approach and attitudes of companies to this issue. Unless both sides take stock of the problems and engage in remedial actions investment opportunities which could be beneficial to both parties will either be unnecessarily delayed or missed altogether.

PART II

COUNTRY STUDIES

PART II
COUNTRY STUDIES

6 ARGENTINA

6.1 Country Summary

Although Argentina is a relatively prosperous 'developing country', with a per capita income of $2,130 in 1985,[1] the course of its gas development can give valuable insights to the problems and pitfalls that occur elsewhere. Two particular facets of gas are highlighted in the Argentinian example: the peculiarly political nature of gas pipelines, and the problems in raising finance for gas infrastructural investment. These two factors put together have meant that in spite of vast reserves of natural gas, and unsatisfied market demand, gas schemes have been subject to endless delays. All the major problems in Argentina focus around the transmission of gas: not on its discovery, nor on field development, nor on the fostering of new markets, as in other countries. And yet these problems alone have been able to halt progress for a number of years.

The potential for gas in Argentina is enormous, and has been likened to the agricultural frontier in that country in the nineteenth century. A large market has already been developed, led by the industrial sector. This is followed by the power sector and residential markets which, unusually, consume equal amounts of gas. Undoubtedly considerable potential still remains, especially in the power sector, which has historically been dominated by hydro schemes. But wider issues of national debt and macroeconomic problems have occupied the forefront of policy-makers' minds. In recent years the fragility of a newly democratized regime has also made decision-taking more difficult than it might otherwise have been; especially when the votes of province Governors are at stake, each anxious for the industrial development of his own province to be promoted. These same Governors have recently made their voices heard by demanding a share in the management of the two state companies, Yacimientos Petrolíferos Fiscales (YPF) and Gas del Estado (GdE), and the right to be consulted on all energy planning.

[1] World Bank, *World Development Report*, 1987.

Political complications have meant that the vast gas reserves discovered in Loma de la Lata in 1978 have still not been fully opened up, despite frequent shortfalls of gas supplies to industrial and residential consumers in the city of Buenos Aires during the winter months. The only progress in nine years has been the construction of the Central West pipeline by Cogasco, which in itself has caused tremendous problems for other schemes.

The pipeline built in 1972 between Argentina and Bolivia exemplifies some of the political problems involved in such infrastructure projects. After it had been built huge indigenous gas reserves were discovered near the pipeline route in Argentina; but a twenty-year contract committed the government to continued imports, which use up the pipeline capacity and prevent transmission of Argentina's gas to its consuming centres. At the same time the price for Bolivian gas became quite out of line with its value to the Argentine economy. But the dependence of the Bolivian economy on its gas exports effectively removed Argentina's bargaining power. Argentina is still committed to buying gas that it does not need, at a price well above normal international levels. These lessons have been well learned by Argentina's other neighbour, Brazil, which is now extremely hesitant about committing itself to imports from Bolivia or Argentina. The long-term commitment of partners in a gas project can thus cause problems to two governments dealing with each other, just as it can to an oil company dealing with a government. It is impossible to predict all the eventualities which may occur over the lifetime of a twenty-year contract. A high degree of trust and flexible contract terms (especially on pricing) are essential.

On the financial front too, Argentina provides an interesting case-study. Raising capital for any large project in Argentina is far from easy, given the country's heavy indebtedness. There is indeed a very generalized hesitation over investing anywhere in Latin America, which reaches the point of a virtual freeze on funds from US banks. Despite this problem however Argentina led the way in experimenting with a new technique for raising private capital for pipeline projects, with the Cogasco scheme. Unfortunately this suffered badly from the political turn of events, with the South Atlantic dispute and subsequent change in government leading to a re-evaluation of the original agreement. The ensuing dispute not only inhibited further investments within Argentina itself, but also removed the hope of carrying out similar schemes elsewhere. Again, the question must be whether any useful lessons can be learned from this experience, that could point the way forward in other cases.

6.2 The Energy Balance in Argentina

Gas penetration of the energy market in Argentina is now at about 32 per cent, having risen fast from a level of 23 per cent in 1980 and just 3 per cent in 1950.[2] Meanwhile the share of oil products fell from 71 per cent in 1970 to 45 per cent in 1985. The government is aiming to increase the share of gas to as much as 50 per cent by the year 2000 (compared to a current penetration of just 23 per cent in the UK). This policy makes much sense in view of the country's natural resources. In 1984 the ratio of petroleum:gas consumption was about 2:1, while the ratio of resources was 1:2. The consumption of different energy types in 1985 is shown in Table 6.1. Imports represented only a very small proportion of the total, at 7.6 per cent, and consisted predominantly of gas from Bolivia (4.4 per cent).[3]

Table 6.1: Energy Consumption in Argentina. 1985. Thousand Tons of Oil Equivalent.

	Volume	%
Oil Products	19,838	45.3
Natural Gas	13,989	31.9
Hydro	5,872	13.4
Biomass	1,915	4.4
Nuclear	1,373	3.1
Coal	833	1.9
Total	43,820	100.0

6.3 Gas Reserves, Production and Infrastructure

(*a*) *Reserves and Production.* Proven gas reserves in 1986 amounted to 24 tcf, and were found in five main basins. The North-west basin lies near the Bolivian border, including the town of Campo Durán. The Cuyo basin is near the western border with Chile, on the same latitude as Buenos Aires. South of this lies the Neuquén basin (to the west of Bahía Blanca). Moving further south again the Gulf of San Jorge basin lies in Patagonia. The Austral basin covers the extreme southern area of Argentina, including Tierra del Fuego. Of these, Neuquén is much the largest, with four times the proven reserves of the next most promising area, Austral.

[2] Most data in this section is taken from Instituto Argentino del Petróleo, 1987.
[3] Ministerio de Obras, 1985, p. 48.

Table 6.2: Argentine Gas Reserves by Region. 1986. Billion Cubic Feet.

North-west	3,866
Cuyo	81
Neuquén	15,452
San Jorge	1,321
Austral	3,344
Total	24,064

Source: Instituto Argentina del Petróleo, *Argentina 1987*, p. 39.

A significant feature is that a large proportion of the gas is associated with oil, and each basin contains both oilfields and gas fields. Neuquén has both; gas/condensate fields predominate in the North-west and Austral basins, and oilfields predominate in the Cuyo and San Jorge areas. The gas-to-oil ratio is increasing quite rapidly however, rising by an average annual rate of 5 per cent for 1970–83 and as much as 9 per cent for the period 1981–3.[4] While Argentina is still self-sufficient in oil most of the fields are old, and production is declining quite rapidly. Furthermore, the large quantity of associated gas in some of the oilfields means that flaring will be very high unless facilities for marketing the gas are available. In the past large quantities of this associated gas have in fact been flared, in response to the demand for oil and the lack of any means for transporting gas. During 1986 100 bcf of gas was flared, equivalent to 14 per cent of the total production of 695 bcf.[5] The Energy Secretary estimated that 15 per cent of production would again be flared in 1987.[6] The government has attempted to put limits on this practice, with the result that some oil wells have had to be shut down.

In 1986 the government brought out a fifteen-year Energy Plan, which aims to greatly increase use of both gas and hydroelectricity, at the expense of oil. An annual increase of 9.3 per cent in gas consumption is planned for the five years 1986–90. It is estimated that an annual growth rate of 5.2 per cent over that period can be achieved through a reduction in flaring and venting of gas. Output per field is planned to increase from 1986 levels as shown in Table 6.3.

In addition to reduced flaring considerable exploration will be required. The government estimates that reserves will need to be

[4] World Bank, internal document.
[5] Cedigaz, *Le Gaz Natural dans le Monde en 1986*.
[6] *Platt's Oilgram News*, 16 April 1987.

Hydrocarbon Basin
Major Gas Pipeline

Table 6.3: Argentine Gas Production: Actual Level in 1986 and Planned Level in 2000. Billion Cubic Feet per Annum.

	1986	*2000*
North-west	74.5	415.0
Cuyo	3.5	0.9
Neuquén	311.5	642.0
San Jorge	60.0	83.3
Austral	199.8	281.6
	649.3	1,422.8

Source: *Platt's Oilgram News*, 28 August 1986.

increased from today's 24 tcf to nearer 43 tcf by 1995. In other words new gas reserves of 19 tcf need to be found during that period, to allow for planned production levels. While this is an ambitious objective it is true that in the past most exploration has taken place in areas thought to have low gas-to-oil ratios, in an effort to avoid the problems of getting gas to its market.

(*b*) *Gas Imports.* In 1985 Argentina imported 78 bcf of gas from Bolivia.[7] While local resources could easily supply demand for gas these imports continue, due to a combination of historical circumstances. In the early 1970s Argentina had already made considerable investments in gas: the major pipelines existed, and the Buenos Aires grid had been rebuilt in 1970. Finds in the north however were beginning to be disappointing, leaving the Northern pipeline running way under capacity. Supplies to Buenos Aires, both for industry and for households, began to look uncertain. The government therefore signed a twenty-year agreement with Bolivia to import gas. The infrastructural requirements were minimal, since an extension of the existing Northern pipeline would suffice. And the initial price was very reasonable. The plan thus made very good sense at the time.

The great drawback was the length of the contract – a necessary part of most gas schemes, given the high capital investments involved. As the price of oil rose during the 1970s the gas price was renegotiated in favour of Bolivia on several occasions. But shortly afterwards the Argentinians discovered the gas fields in Salta province on the Argentine side of the border with Bolivia, which could have fed the pipeline with domestic gas. The fall in the oil price could probably have been accommodated in some measure (since a rise had been) were it not

---

[7] Ministerio de Obras, 1985.

that in the mean time Bolivia's economic situation had become unbelievably bad. By 1984 the export of gas to Argentina was Bolivia's main legitimate export. Even to cut the price, let alone to terminate the contract, was politically impossible for Argentina. The high price being paid for the Bolivian gas had enormous impact on the financial state of Gas del Estado, and the economy in general. For example, in the summer of 1986 GdE was paying $4.20/mbtu, compared to $0.78/mbtu (including royalties) for gas from YPF. Later in 1986 these terms were improved slightly, to $3.70/mbtu to be paid for half in products, but this is still above international prices, and fresh negotiations are under way. For political reasons the government's hands are more or less tied however.

(*c*) *Infrastructure.* There is an extensive network of gas pipelines in Argentina, reaching from the southern tip of the continent to Campo Durán in the far north of the country. There are four major systems, serving the different gas fields, as shown in Table 6.4.

Nevertheless large quantities of gas are still flared, while the demand for gas in Buenos Aires and other cities cannot be satisfied during winter months. The current Energy Plan includes ambitious plans for expanding and extending the network to try and solve this dilemma, but finance presents a severe constraint to the realization of those plans.

The 1986 Energy Plan contained proposals to increase gas transmission capacity by 60 per cent, to reach 3,715 mcf/d by the year 2000. This would include expansion in the capacity of the existing Central West line, from 353 to 636 mcf/d; expansion of the Northern pipeline, which runs from Campo Durán to Rosario, from 335 to 477 mcf/d; the construction of a new line from Neuquén; a line to the eastern provinces and possibly to Uruguay and Brazil, crossing the Parana river; and a bevy of small regional lines, connecting gas fields

Table 6.4: Argentine Gas Pipeline Network.

Pipeline	Gas Field	Length (miles)	Capacity (mcf/d)
North	North-west Basin	1,743	335
Central West	Cuyo/Neuquén	1,017	353
West	Neuquén	1,219	381
South	Austral/San Jorge	2,675	448

Source: Instituto Argentino del Petróleo, *Argentina 1987*, p. 41.

and the main trunk lines to nearby population centres. While some funding towards these projects is expected from sources such as the Inter-American Development Bank (IADB) and the World Bank there will also be a need to raise considerable sums from elsewhere.

In the past all pipelines fell under the monopoly of GdE; but its lack of cash meant that much-needed expansion was delayed. As a result private capital was for the first time invested in the system, with the construction of the Central West line in the late 1970s. While this was not altogether a success, as described below, it did set a precedent for the general principle of private involvement. Even under that scheme ownership of the line eventually was to revert to the state. But an announcement was made in November 1986 that in future private investors will be allowed to build, operate and own gas lines to serve small and medium-sized rural communities.

Financial problems over pipelines have been further complicated by political concerns. This is most clearly illustrated in the cases of the Cogasco and the planned Neuquén lines. The full stories of these two projects are therefore given, as they illustrate well the problems that can arise around gas infrastructure projects in any country.

Cogasco. This Dutch-led consortium was formed to build and run the Central West pipeline. Cogasco would raise the required finance without Argentine involvement. Gas del Estado would then pay Cogasco a tariff for the use of the line over twenty years, at the end of which period the line would revert to GdE. The tariff would be paid 5 per cent in local currency and the remainder in foreign exchange. (The 5 per cent was not necessarily intended to cover all local costs, since the company was happy to use some of its dollars earned in Argentina within the country, rather than struggle to obtain permission to remit them all to the Netherlands.) The Argentine government guaranteed this tariff payment (which was linked to LIBOR), but *not* the loan. The investment was thus on a non-recourse basis.

Cogasco raised the capital by borrowing from its parent company, Royal Boskalis, from the banks (including British banks), and others. There was also a large loan from Dutch banks with export credit guarantees from the Dutch government. This loan was tied to the exchange rate between the dollar and the guilder, which left scope for gains or losses according to variations in the currency values. In the event the Dutch would have made very large gains if tariff payments had gone according to course. This has received unfavourable comment in Argentina.

The 1,125-mile line was built, and came on stream in 1981. But in 1982, at the time of the South Atlantic conflict, GdE stopped paying the tariff, except for the local component. Given the exchange rate at which the deal was set up, and the rise in LIBOR, GdE would by 1982 have been paying no less than 60 per cent of its income to Cogasco. But in any case at this point Argentina was paying almost none of its debts, and certainly no British debts. The small British involvement was a more than adequate pretext for not paying what could not be paid in any case.

The gas continued to flow (presumably for fear of nationalization), but Cogasco soon met with further difficulties. The 5 per cent payment in local currency had never been intended to cover all local costs, but the manner of indexing it meant a rapid deterioration in its value. Cogasco had to borrow heavily in Argentina and began to put pressure on the government to improve the situation. The military government was unresponsive, although a new pressure was now developing: the need to move to the second stage of the Cogasco project and install compressors to increase the capacity of the line. When Alfonsin took over power, the new Energy Secretary brushed aside the problem and set to work on an alternative pipeline, Neuquén–Bahía Blanca–Buenos Aires, in order to solve the political need to fill the growing winter gas deficit in Buenos Aires. The situation between Cogasco and the government was to worsen still further. In 1984 in the context of trying to solve the general debt problem the Central Bank demanded a declaration of all private debts. Cogasco declared $1.2 billion; the Bank inspectors could find only $900 million. Cogasco claimed the rest was accounted for by un-invoiced 'royalties, fees etc.' The resulting accusations of bribery led to a further deterioration in the atmosphere. The huge speculative profits which the Dutch banks stood to make due to currency fluctuations added to the controversy.

Meanwhile in the course of 1985 outside pressure mounted to resolve the issue – particularly via the Paris Club, where banks and international agencies took the line that resolution of this issue was necessary to re-establish credit-worthiness. Both the World Bank and the IADB avoided making further loans to Argentina for energy projects until it was resolved. In March 1986 the Dutch banks imperilled Argentina's debt rescheduling by threatening to withdraw. Meanwhile Royal Boskalis was going bankrupt, and was rescued by a consortium of Dutch banks.

Various solutions to the problem began to be explored under the new Energy Secretary, Jorge Lapena, who halted the tendering for the alternative pipeline. A solution was finally reached at the end of 1986, whereby the Dutch government assumed Cogasco's $1.017 billion debt. Ownership of the line reverted immediately to the Argentinian government, which also undertook to repay the loan to the Netherlands. The expansion of the line will not now be paid for by Cogasco, as had been originally agreed. But tenders for work to expand throughput from 353 to 636 mcf/d are being taken. Costs are estimated at $200 million.

The major problem with the scheme, from its inception, was the uneasy relationship between a private pipeline owner and a state-controlled industry. Both supply and demand for gas were tightly under government control, whilst the pipeline owners wanted to run their operation according to commercial criteria. While the investment was sound in the sense that gas throughput could earn an adequate return, the intervention of political events brought a halt to payments. The outbreak of a war is a relatively uncommon and extreme manifestation of 'political risk'; but nevertheless each vindication of investors' nervousness adds disproportionately to future hesitation. The Cogasco saga has thus been very unfortunate, since it has put back the chances of raising private capital for gas infrastructure elsewhere as well as in Argentina itself. For future schemes Argentina is now looking primarily to the World Bank and IADB for investment funds.

Neuquén. A new pipeline from the Neuquén gas field has been under discussion for a number of years. While there is clearly a need for such a line, given projected demand for gas, the issue has been at the centre of several political battles. The arguments have been over when it should be built, in relation to other lines; and over its route. The timing problem has been a relatively simple case of playing off one negotiator against another, since the immediate options were between building this new line and expanding the capacity of the existing Cogasco line; while both courses would eventually be necessary a choice had to be made over which came first. When Cogasco negotiations were going badly the Energy Secretary threatened to act quickly on constructing the Neuquén line, thus lessening Cogasco's bargaining hand. In 1986 the incoming Energy Secretary, anxious to solve the dispute with Cogasco, once again postponed the Neuquén line. Relations with the Dutch did improve and solutions gradually emerged. The whole process of negotiation over Cogasco still took many months however, and political pressure from unsatisfied customers was growing. By July 1986 it was felt that no further delay could be tolerated, and the government decided to award a contract for the new line. They did this without going to public tender, in order to make up for lost time. It has in fact been awarded to an Argentine/Mexican consortium, amidst widespread concern about this hasty procedure.

Other potential bidders were understandably put out.

The issue of the pipeline's route has been even more fraught with political complications, since it has aroused various regional interests in Argentina. The reason for this is that the chosen route will have an enormous impact on the location of future industrialization. The Governor of Neuquén has energetically championed the cause of his state, in which he would like to see a petrochemical industry. In objective terms this does not make much economic sense, not least because there is no port in Neuquén. But his arguments are not without political weight; for Alfonsin's government does not have a secure majority, and Neuquén has two independent senators who wield considerable power. Meanwhile regional interests in Santa Fe and in Salta are anxious to secure new fertilizer plants for their areas. The province of Buenos Aires is also pleading a special cause. And the favoured route on economic grounds is for the line to be routed via Bahía Blanca, where there is already a large petrochemical industry in need of more feedstock. These opposing claims posed a considerable balancing trick for the government, which did not yet feel itself to be very secure.

The decision was taken however, that it should go from Neuquén to Bahía Blanca, and thence to Buenos Aires. The plan of the Argentine/ Mexican consortium is to first build the 330-mile stretch to Bahía Blanca, to be followed by the line to Buenos Aires (394 miles). An additional section will encircle Buenos Aires (90 miles), to supply the industrial, commercial and residential markets in the region. The contract has been awarded, and the scheme is due to come on stream in 1988. The total cost is now put at $497 million; $195 million from Mexican banks and industrial sources, $40 million from Siderexport of Italy; $40 million from Japan's Eximbank, and $17 million from Mitsubishi and Mitsui. Much of this will be in the form of supplier credits, especially for linepipe. An interesting suggestion has been that Buenos Aires customers should help to finance it, through advance payment for their connections; but this could only cover a small proportion of the total project costs. The balance will be raised locally through the issue of 'gas bonds', to be repaid by Gas del Estado from the proceeds of their increased sales.

6.4 The Government Sector

The government sector is crucial to the history of gas utilization in Argentina, since there has been a long history of strong opposition to foreign companies operating in the country. Consequently the national oil company, YPF, produces around 80 per cent of all gas from

Argentinian fields. Many of the problems in developing the gas market also stem from political power struggles, which find their clearest expression in the relations between the various government agencies.

(*a*) *Energy Secretariat*. This falls under the broad umbrella of the Ministry of Public Works. It is divided into four operational sub-secretariats: Electricity, Fuels, Planning, and Relations with Enterprises. The sub-secretariat of Fuels, amongst other things, is in charge of supervising the three state-owned enterprises for oil, gas and coal resources respectively: Yacimientos Petrolíferos Fiscales (YPF), Gas del Estado (GdE) and Yacimientos Carboníferos Fiscales (YCF). This supervisory task is very difficult to carry out however, given a very small staff in the Energy Secretariat, and the entrenched position of YPF in particular. Nevertheless efforts are being made to curb the power of the national companies. The ambitious fifteen-year Energy Plan which was published in 1986 was produced by the Energy Secretariat.

(*b*) *Yacimientos Petrolíferos Fiscales (YPF)*. YPF is the key agency to gas utilization in Argentina. It is very large, has a long history, and is very powerful. Nevertheless it has an extremely low reputation within the country, and foreign oil companies do not find it an easy partner to work with in joint ventures. Its problems may be summarized under four headings: the 'YPF-mentality', poor management, government-controlled prices, and an appalling financial situation (which is of course a result of the other factors).

The 'mentalidad YPF-eana' is a product of inherited attitudes. Two aspects of this are particularly important to gas development. First there is a long tradition of opposition to the role of the private sector, and particularly foreign capital in oil. The 1985 Houston Plan, which welcomed foreign capital, came direct from President Alfonsin and has had to be imposed upon a sector which developed under a completely contradictory philosophy. This does not make for easy relationships with foreign partners. The second aspect of the inherited mentality is a lack of interest in gas. The company was created to find and produce oil; although gas now constitutes 40 per cent of YPF's production in terms of thermal equivalence it is still seen as a poor relation. YPF's role in gas is also much more limited than in oil, given the existence of Gas del Estado, which is responsible for both transmission and distribution. The growing importance of gas therefore threatens to diminish the political power of YPF, and not unexpectedly this meets with some resistance.

The management structure of YPF has been widely criticized for many years. It has been notable for over-employment, poor quality of management and lack of incentives arising from inadequate promotion structures. Key appointments have tended to be made on political grounds. The pattern has therefore been for talented staff to move into other more challenging companies. A particularly insidious aspect of the management problem has been constant suggestions of large-scale financial irregularities. The government has attempted on several occasions to tackle these problems. For example, there was a plan in 1986 to split it into four separate companies, to try and break down some of the inertia caused by years of poor management. But this was later changed to the creation of three divisions within the company (exploration and production, refining, transport and marketing). This type of strategy has been tried before, but without very much success.

The gas price at the well-head is set by the government, and while it is sufficient to recover production costs it offers no incentive to explore for gas. This has had obvious implications for the balance sheet of the company, but has also contributed to the corporate lack of interest in gas, since it is 'valued' less than oil. There may be some economic justification for maintaining low prices, but the effect on the company is appalling, since between half and two-thirds of the already low price goes in royalties to the provinces. (Royalties are calculated on the basis of fuel oil equivalent, whereas the price is not – it is far lower.) Under these circumstances, when YPF's budget is being squeezed and it is being criticized for its low rate of oil exploration, it naturally concentrates its marginal effort on oil.

The financial position of YPF is dire, and it is under great pressure to rectify this. While there are constant accusations of irregular practices, there are also very good reasons why the deficit has arisen. First the low well-head prices for gas and oil, and high royalties to the provinces, do not provide an adequate income for the services provided by YPF. Secondly it has an inherited problem from the time of the military regime, when it was used as a means of raising foreign loans. Thus money would be borrowed against its balance sheet for other, usually unproductive, uses (a not uncommon practice in Latin America). Thirdly, during 1982 it was forced to accumulate enormous losses and debt when product prices were frozen at a time when inflation peaked at 30 per cent per month. It is very difficult to see how YPF can extricate itself from these problems, given its lack of control over prices. The recent improvement in gas prices paid to producers (a ceiling at 25 per cent instead of 14 per cent of crude oil prices) may help to some extent. But in the mean time the commitment of YPF staff is sapped still further, by the apparently insuperable financial problems of their company.

(c) Gas del Estado (GdE). Gas del Estado was founded in 1946 and is responsible for transmission and distribution of gas. When compared with the gas producer, YPF, it is generally viewed as being efficient and sensible. Its problems come from its smaller size, and consequently from its lack of weight against YPF. This is important when there are differences of opinion between the two companies, especially since YPF tends to feel threatened by the ascendancy of gas over its traditional oil business. This finds its expression in disputes over issues such as which company should be responsible for the treatment and separation steps in gas production (usually carried out by GdE). GdE also has financial problems. These are partly caused by the very high price it is forced to pay for Bolivian gas. The company is effectively being penalized for a political decision not to renege on the government's obligations to its neighbour. The major cause however is low consumer prices for gas. For example in 1983 GdE could cover only about two-thirds of its operating costs from gas sales. In principle the Treasury has agreed to take over the excess payments arising from the Bolivian contract, but in practice only partial compensation has so far been paid. To reform consumer prices is an even more intractable problem.

6.5 The Private Sector

Foreign firms must either be incorporated in Argentina or belong to a consortium which includes a local company. Exxon, Shell, Total, Deminex and Occidental are involved in exploration; Esso and Shell in refining and marketing of oil products in Argentina; Occidental and Amoco in production. There are also several local private companies active in the oil and gas sector, such as Astra, Bridas, Perez Companc and Pluspetrol. Nevertheless it should be remembered that only a very small proportion of the gas produced (about 25 per cent) emanates from the private sector. YPF has the lion's share of this market. Most of the gas produced by foreign companies is associated with their oilfields.

   Three local companies currently produce some gas: Bridas, Perez Companc and Pluspetrol. Amongst the foreign companies present in Argentina there has been little activity on gas. Shell investigated the possibility of selling associated gas from its oil discovery in Magellan, but found it to be uneconomic. More recently an interesting project is taking shape off shore in the Austral basin, under a Total/Deminex/ Bridas consortium. A combination of luck, new methods and a talented exploration team, has recorded a success ratio of ten finds in twenty-six wells drilled. The first stage of development concentrated on the Hidra oilfield, which has a very low gas-to-oil ratio, but negotiations have also taken place over development of the neighbouring Ara gas field, which is scheduled as the next stage of the project. A gas price of 50 per cent of the oil price has been agreed, which is quite acceptable. For the next field, which lies further off shore and is deeper, they have agreed the same gas price, on condition that they receive a better price for their oil.

Companies' interest in oil exploration has remained quite high, despite internationally low oil prices. Shell, Exxon, Amoco and Oxy are all tendering for new concessions. None of the companies is enthusiastic about finding gas however, due to the shortage of transmission capacity.

Corporate existence in Argentina is not seen as being too difficult. While newcomers might experience problems in adapting to the political lobbying, the bureaucracy, and similar local characteristics, most of the companies operating in Argentina have been there for a relatively long time, and know their way round the system. While their activities, and especially their returns, are constrained by state controls over production and distribution, they have established a workable *modus vivendi*.

One major disincentive to foreign companies is the requirement that they always have a local partner. Even local private companies behave according to a different ethos than that of the foreign oil companies – they look for safe and rapid returns, and are not risk takers. This difference in approach can make joint projects very difficult. Even more of a problem is posed by YPF. The companies find YPF over-bureaucratic and badly managed, and often feel that it does not really welcome private foreign capital. It is felt that YPF keeps the most promising exploration areas for itself, with private companies only having access to second-best areas. It also seems that YPF could lobby the government more effectively on behalf of private interests if it so chose, and this does not foster goodwill. Recent changes in the model contract have greatly reduced the power of YPF in joint ventures however.

6.6 Gas Policy, Contracts and Pricing

(*a*) *The Model Contract.* In August 1986 the government issued a Model Contract for the exploration, development and exploitation of oil and gas. While intended to encourage greater activity by both foreign and local companies its original provisions caused some disquiet amongst foreign companies. In May 1987 President Raul Alfonsin issued an executive decree which altered some clauses, and gave foreign companies greater flexibility in investment decision-making. In particular, the potential for interference by YPF was greatly reduced.

For example under the original terms YPF had the exclusive right to declare a prospect commercial. Companies were worried that YPF might then be tempted to declare an area 'non-commercial' in order to be able to exploit it itself at a later date. The decision now rests with the private operator and partners: it must normally be taken within one year of discovery. YPF will be given one opportunity to associate during the production phase, for a 15–50 per cent share. It will be required to fully compensate the private partners for exploration and development costs previously incurred.

Payment and pricing provisions were also changed, following protests from the private companies. Since all production has to be sold to YPF the companies are heavily dependent on YPF's ability to pay in foreign exchange. The original Model Contract allowed for payment either in crude oil or oil products if sufficient currency was not available. While international f.o.b. prices were stipulated companies did not want to carry the risk of payment in unspecified products for which the world market was unpredictable. This provision has now been eliminated, and the only alternative to dollar payments is crude oil.

(b) *Pricing*. Consumer prices for gas in Argentina are low. They are broadly equivalent to the long-run marginal costs of gas production and distribution, but are below the international prices for alternative fuels, and are insufficient to pay for urgently needed infrastructure investments. In addition prices to residential consumers are relatively low while those to industry are high, whereas the reverse should be true on economic grounds.

Prices paid to the producers are also a highly contentious issue. Under the new Model Contract the price paid to the oil companies was originally set at 14 per cent of that for Arabian Light, or the price paid by GdE to YPF, whichever was the higher. This was subsequently amended to give a minimum price of 17 per cent of the international price, with a maximum of 25 per cent. However a very large proportion of this price goes towards royalties to the provincial governments. As of August 1986 the companies then producing gas were receiving only about $0.20/mbtu of gas, net of royalties. This especially affects YPF, which produces 75 per cent of all gas. It contributes to YPF's deficits, and reduces all the companies' interest in producing gas.

Companies' lack of interest in gas is not an immediate problem in Argentina, since known reserves will provide more than enough gas for planned future requirements and pipeline capacity. There is therefore no need to encourage companies to look specifically for gas. Both the Argentine Energy Secretariat and outside bodies such as the World Bank are agreed on this. But the issue is not that simple. For if the companies have reason to think they may find more gas than oil, they will effectively be discouraged from looking for oil, and the government definitely does not want this to happen. Hence the readiness of President Alfonsin to amend the Model Contract in the face of company disquiet.

The government has also responded by pursuing a very flexible policy, which helps to meet the needs of the foreign companies. Thus rather higher prices for oil have been agreed than would normally be expected, while still offering a low or nil price for associated gas. While companies might prefer full commercial pricing for the gas, this solution undoubtedly helps in some situations.

Finally, the issue of Bolivian gas imports should be mentioned. As described above, this contract was drawn up at a time when Argentina badly needed foreign supplies of gas. The price agreed upon was therefore quite high ($4.50/mcf); and enormously higher than the eventual costs of producing indigenous gas. In May 1986 the Argentinians succeeded in reducing the price to $3.70/mbtu, but this was still well above the international level, which was nearer to $2.50/mbtu. They also agreed that payment should be made 60 per cent in cash and 40 per cent in products; but they are now trying to reduce the price further, in return for payment entirely in cash.

6.7 Markets

Gas is very well established as an energy source in Argentina; the problem is one of transporting sufficient gas to consumers, to meet the unsatisfied demand. While the gas fields are well distributed through the country, the gas market is overwhelmingly concentrated in the province of Buenos Aires. A well-developed town grid was constructed in Buenos Aires in the nineteenth century, and has now been converted from town gas to natural gas; and there is a substantial industrial base in which gas could be more widely used. Demand for gas is currently growing at 15–18 per cent per annum, but there is a bottleneck in the shortage of pipeline capacity. This is the major problem facing the gas industry in Argentina, as discussed in Section 6.3.

The growth of the different markets for gas is shown in Table 6.5. The pattern of gas usage is very different to that in most developing countries. Most marked is the fact that the industrial market far outstrips that of the power sector. Even the residential sector consumes slightly more gas than the power sector – despite the fact that 35 per cent of power is generated from gas. This pattern appears to be on the way towards that found in Western industrialized countries, where typically the residential market accounts for the largest share, followed by the industrial, with the power sector taking a much smaller share. The issue of what proportion of electric power should be generated from gas and hydro respectively is still hotly debated within the country.

Table 6.5: Argentine Natural Gas Consumption by Sector. 1973–85. Billion Cubic Feet.

	1973	1975	1977	1979	1981	1983	1985
Residential	45.4	53.8	57.6	66.3	81.6	109.3	122.3
Industrial	111.4	127.7	144.7	144.8	135.7	170.3	183.5
Power	55.1	64.3	61.8	62.4	83.1	109.4	118.6
Commerce/ Government	12.9	15.2	17.9	21.7	30.3	42.4	42.8
Others	4.1	2.2	2.5	–	–	–	–
	228.9	263.2	284.5	295.2	330.7	431.4	467.2

Source: Gas del Estado, *Boletín Estadístico Anual*, 1985, p. 45.

(a) *Power.* In 1986 42 per cent of power was generated from hydro-electric plants, 28 per cent from gas, 14 per cent from nuclear, 13 per cent from fuel oil, and 3 per cent from coal.[8] Thus, while gas did contribute a large amount towards this sector, an even larger share was taken by hydro. The decision to favour hydro so heavily was taken before the large Neuquén gas field was discovered.

This decision was based on cost comparisons using the opportunity cost for gas at the very high Bolivian border price. Once the Neuquén field was discovered, and the North-west fields were proved to be much larger than previously imagined, the same policy team which had recommended hydro schemes reversed its decision, in favour of gas power stations. These individuals were now out of power, but their successors in the Secretariat of Energy also accepted their argu-

[8] World Bank, direct communication.

ments. To implement such a change in policy proved virtually impossible however. The planners met with political difficulties, caused by the lobbying power of the different groups involved. They also encountered the high cost of stopping a project once it is under way. For all such public works contracts had heavy financial penalties built into them. It is clear that such costs radically affected the room for manoeuvre in response to the change in relative prices, which had clearly occurred. (The rise in interest rates also affected badly the highly capital-intensive hydroelectric projects.)

The recent Energy Plan, designed to influence policy for the next fifteen years, gives increased preference to use of gas in the power sector, although hydro is still favoured as a renewable resource. Some commentators would like to see a greater emphasis on gas for power generation. The major limiting factor in this, as in other sectors, is the capacity of the gas transmission system. Existing thermal stations tend to be dual-fuelled, since they are only able to use gas in the summer months: in the winter residential and industrial consumers are given preference over power stations, and supplies to the latter are curtailed. Gas consumption in the power sector could therefore increase substantially once the pipeline capacity is improved, even without the need for any investment in additional power stations.

(*b*) *Industry*. A high proportion of industries in the Buenos Aires area already use gas; but there is still much unsatisfied demand. Hence the planned ring pipeline around the city, which should form part of the Neuquén project when it finally comes to fruition. It is interesting to note that it is assumed that the consumers will be prepared to put up capital in advance, to help finance such a line. This is a good indication of the level of acceptance for gas as a fuel.

(*c*) *Residential*. The penetration of gas into the residential market is very marked, and is a reflection of the relatively cold winters in the main cities, coupled with the existence of a sizeable wealthy class. This is a major distinguishing factor between Argentina and other developing countries looked at in this study. The growth of this market owes much to the political benefits that can be gained through cultivating support of a class with economic power. Prices for domestic gas in wealthy suburbs are kept artificially low, while poorer sections of the community pay relatively more for LPG and kerosine. In 1984 the ski resort of Bariloche was connected to a gas supply, while it would have been much cheaper to take the lines to poorer suburbs within Buenos Aires. It has also proved to be a political liability for the Energy Secretariat however. It is particularly difficult to substantially raise gas prices when so much consumption is accounted for by thousands of individual households, each of which has its voters.

(*d*) *Transport.* A programme is under way to convert 30,000 vehicles (mostly taxis) in the first instance to run on CNG, to be extended later to at least 100,000 vehicles. Finance for the project is coming partly from the World Bank, to cover the cost of twenty filling stations ($5.8 million) and technical assistance ($2.1 million) on the planning, implementation and management of the programme.

(*e*) *Fertilizers.* Installed capacity in Argentina is 82,000 t/yr of ammonia and 100,000 t/yr of urea, consuming 12 mcf/d of natural gas. Recent projects to expand this capacity have included the award of contracts in September 1986 for construction of a 100,000 t/yr ammonia/urea plant at the Neuquén field. Following that, bids have also been invited for a major fertilizer complex, to produce 350,000 t/yr of ammonia. The location for this plant is still open however, and the different provinces are in competition to win it. 80 per cent of the production from this plant would be destined for the domestic market. A further 100,000 t/yr ammonia/urea plant is projected for Salta.

(*f*) *Petrochemicals.* Argentina already has two methanol plants, producing a total of 36,000 t/yr, and consuming 3 mcf/d of gas. A number of additional projects were being planned early in 1986; but by the end of the year all activity on these had been suspended, due to the low international prices for methanol. The plans had included a plant at Río Grande, Tierra del Fuego, with a capacity of 680,000 t/yr; this $300 million contract had been awarded to the Spanish firm Tecnicas Reunidas. A second plant was due to be constructed at San Lorenzo, near Buenos Aires, in which GdE would have had a 19 per cent interest. This was to have had a capacity of 150,000 t/yr, and would have cost $63 million.

There is already a flourishing petrochemical industry, centred on Buenos Aires and Bahía Blanca. However the ENI group is currently carrying out feasibility studies for a further five plants, requiring an investment of $300 million, aimed at the export market.

(*g*) *Regional Markets.* An unusual amount of discussion has taken place in Latin America over the possible integration of gas resources between countries. As far back as the 1960s the general manager of Gas del Estado, Esteban Perez, a man of exceptional vision, saw gas as the key instrument to regional integration. He first envisaged the network which is still dreamed of. Bolivia has exported gas to Argentina since 1972; and future links are planned between Argentina, Uruguay, Brazil and Chile. While these are all beset with potential political pitfalls, and the National Energy Plan does not therefore include them in its projections, meetings between the various heads of government continue to have gas pipelines high on the agenda. Because of its huge proven reserves Argentina is of course seen as the major exporter of gas, despite the fact that at present it is an importer.

Uruguay. There has long been discussion over the possibility of Argentina exporting gas to Uruguay. The most recent scheme was for a 350-mile pipeline from Santa Fe to Montevideo, to carry a minimum of 42 mcf/d. The cost of this was estimated at about $175 million, and the annual income to Argentina at about $100 million. As negotiations with Brazil have progressed however the logical course seems to be to combine the two schemes, with a pipeline passing through Uruguay on the way to Brazil. The Uruguayan President is now lobbying in favour of this, which would obviously reduce costs. The political implications are even more complicated than for a bipartite agreement however, and no concrete progress seems yet to have occurred.

Brazil. The first official expression of interest in exporting gas to Brazil came in 1980, when the Brazilian President visited Buenos Aires. Feasibility studies were commissioned and meetings scheduled between Petrobrás and Gas del Estado. Even at this very early stage however resistance was felt to the idea from a number of quarters. On the Brazilian side there were two sources of discontent. The military were hesitant at becoming dependent on Argentina for an important energy source. Even more importantly, Petrobrás feared the loss of its traditional market, and developed the notion that the country lacked a 'gas culture'. Gas supplied only about 0.3 per cent of the energy balance: not only did the infrastructure not exist, but potential customers had no experience with the fuel. On the Argentine side, there was extreme uncertainty over the issue of security of supply. Over and above the understandable desire of bureaucrats to hoard what is theirs, there was concern over the winter gas supply to Buenos Aires. This led to the difficult proposal that Argentina might supply substantially less gas during the four months of winter. The prices proposed at this stage were far apart – $2.50/mbtu as against $5.00/mbtu.

Negotiations came to a standstill, and the scheme faded from public view for a number of years. During this period the fall in the international oil price and Brazil's discovery of further oil and gas of its own, diminished the appeal of importing gas. Petrobrás then became concerned that investment in infrastructure to bring gas from Argentina might pre-empt development of domestic gas reserves. High interest rates also made the financial burden of such a scheme quite prohibitive.

In 1986 however the scheme was revived, as part of an overall *rapprochement* between the two relatively new democratic regimes, under which a Protocol of Agreement was signed in July. The governments are actively seeking joint projects, in an effort to strengthen political and trade relations. For example they are also planning a joint hydroelectric scheme on their borders, and hoping to integrate their electricity transmission systems. The new gas proposal is more modest than the earlier plan to pipe gas as far as Sao Paulo. Since those days a sizeable market has developed in the southern province of Brazil, and the plan is to build a pipeline from Campo Durán to Porto Alegre in this province, a distance of 812 miles. A joint Petrobrás/YPF/GdE study has put the cost at $230 million for a basic system. It is judged to be economically viable with a minimum throughput of 53 mcf/d, at a gas price of $2.50/mbtu. The targeted market is in power plants, a petrochemical complex, and about 120 manufacturers. There are no plans to pipe the gas to households. Estimates are that the market could expand to as much as 106 mcf/d.

While technical studies and costings are under way, no firm agreement has yet been reached over any of the important details, most crucially that of pricing. Even if this can be settled there will still be the matter of raising the necessary finance. But nevertheless prospects of real progress are better than they have been for some time.

Chile. Chile has discovered gas in Tierra del Fuego, but has no pipeline to take it the long distance to its industrial centres in the north of the country. A logical plan has been put forward, whereby Chilean gas in Tierra del Fuego would pass through the Argentinian pipeline which already exists, while a new line could cross the Andes from Neuquén to Santiago to supply the Chilean market. While this makes sense in abstract terms, it is most unlikely to come about, due to political and territorial problems. The planned construction of a Chilean methanol plant in Tierra del Fuego will effectively remove the need for the scheme, by providing an outlet for Chile's gas.

6.8 Finance

The overall macroeconomic situation of the country is a major handicap to financing gas projects in Argentina. Four problems arise:

(a) cut-backs in government spending; (b) the reluctance of foreign private investors and banks to commit capital to a country with a bad repayment record; (c) problems in raising local private capital; and (d) delays and difficulties in raising funds from multilateral aid agencies. Over and above these basic problems finance is usually offered as part of a 'package' which includes all kinds of other political and economic strings, some of which may be acceptable to the government and others not.

Government investment in the hydrocarbon sector has traditionally been of great importance, and has accounted for as much as 40 per cent of total public investment. The current austerity programme, to reduce the public deficit, must therefore have a great impact on the sector. As well as cut-backs in direct investment in gas projects, the restrictions on new projects for electricity generation and refinery conversion have an indirect effect, by reducing the potential outlets for gas. It has therefore become vital to attract private funds into the gas sector.

Foreign banks have shown no willingness to commit funds to Argentina, over and above the obligatory new funds negotiated as part of the rescheduling of the country's external debt. Private companies have offered capital investment in projects in which they are directly involved, such as methanol plants. But the outstanding problem in Argentina is the need to invest in additional gas infrastructure. The Cogasco project was originally heralded as a breakthrough in raising private capital for such schemes. But of course following the fiasco with this scheme other investors have not been forthcoming with similar funds. The situation should ease slightly now that the Cogasco problem has been settled; but it is unlikely that others will rush into similar infrastructure schemes.

Ironically, raising local private capital proves to be even more difficult than raising support from foreign investors, thanks to the capital flight from Argentina that has occurred over the last decades. Consequently, where there is a choice between two schemes, one using local components and construction firms, and the other using foreign equipment and hence needing foreign exchange, the government is likely to favour the latter. The problem of local funding has also been cited as a reason for the delay in expanding the Northern gas pipeline. Although the IADB promised funds for this project some time ago the government has been unable to take up the offer because it cannot raise the local currency component.

International aid agencies have provided large sums towards the development of the gas industry in Argentina, but there are a number of criticisms of the way in which this has been done. The agencies are often bureaucratic and slow, drawing out even further negotiation processes which are inevitably slow in the first place. They also have rather stereotyped procedures and perceptions, raising the same issues for argument over and over again, such as increasing the prices charged by public utilities.

Finally, political issues are always intimately bound up with the financial. For example the decision to use foreign suppliers, because foreign exchange is more easily come by than local currency, is bound to be unpopular with the local population and state governments. Similarly, offers of foreign finance usually come with strings attached: the US Export–Import Bank threatens to withdraw its financing if a bid from the USSR is accepted; the Mexican government offers to take quantities of Argentinian wheat, etc. The problem is thus not just one of raising the capital, but also of carefully balancing the various interests involved. This can be particularly delicate for a relatively new democratic government, which is trying to build its constituency.

Sources

Gas del Estado, *Boletín Estadístico Anual*, 1985.

Guadagni, A. A., *Decisiones Energeticas para el Futuro*, Instituto Torcuato di Tella, September 1986.

Instituto Argentino del Petróleo, *Argentina 1987*, brochure prepared for 12th World Petroleum Congress.

Ministerio de Obras y Servicio Publicos, Secretaria de Energía, *Anuaria de Combustibles*, 1985.

World Bank, unpublished papers.

In addition to these publications, and those cited in the footnotes, valuable information was obtained from the following sources: Chase Manhattan Bank, Gas del Estado, A. A. Guadagni, Ruhrgas, Secretaria de Energía, Shell, YPF and the World Bank.

APPENDIX TO CHAPTER 6

Background Data

Population[a]: 30.5 million
GNP per capita[a]: $2,130
Average annual growth in GNP per capita, 1965–85[a]: 0.2 per cent
GDP[a]: $65,920 million

External public debt[a]: $40,179 million
External public debt as percentage of GNP[a]: 56.4

Total energy consumption[b]: 43.820 mtoe
   Commercial: 96 per cent   Non-commercial: 4 per cent
Commercial energy consumption per capita[b]: 1,445 kgoe
Commercial energy imports[c]: 3.322 mtoe
Commercial energy exports[c]: 3.226 mtoe

Proven Reserves and Production[d]

	Gas	Oil	Coal	Hydro
Proven Reserves	24 tcf	2.3 bbls	171 mtoe	40,000 MW
Production	371 bcf	430,000 b/d	400 kt	5,108 ktoe

Consumption by Sector: Thousand Tons of Oil Equivalent[f]

	Gas	Oil Products	Coal	Total	% Share
Power	3,029	2,655	272	5,956	18
Industry	3,507	1,503	17	5,027	16
Transport	–	10,150	–	10,150	32
Households	3,311	1,353	–	4,664	14
Agriculture	–	1,465	–	1,465	5
Other	2,265	2,539	25	4,829	15
Total	12,112	19,665	314	32,091	
% Share	38	61	1		100

Power Sector

Installed capacity[c]: 11,290 MW
Electricity generation (%)[g]: Hydro=2, Oil=13, Coal=3, Gas=28,
Nuclear=14

Sources

(a) World Bank, *World Development Report*, 1987, pp. 202–237. Figures refer to 1985.
(b) Instituto Argentino del Petróleo, *Argentina 1987*, brochure prepared for 12th World Petroleum Congress. Figures refer to 1985.
(c) Ministerio de Obras y Servicio Publicos, Secretaria de Energía, *Anuaria de Combustibles 1985*. Figures refer to 1985.
(d) Gas and coal reserves taken from (b); gas production and oil reserves and production taken from *BP Statistical Review of World Energy*, 1987. (Figures refer to 1986.) Coal and hydro production taken from (c); hydro potential taken from (e).
(e) World Bank, *The Energy Transition in Developing Countries*. 1983. Figures refer to 1980.
(f) World Bank, internal reports and direct communication. Figures refer to 1983.
(g) As (f). Figures refer to 1986.

7 EGYPT

7.1 Country Summary

Egypt has large gas reserves (about 9 tcf proven), but has taken a long time to enact gas legislation specifically designed to encourage exploration. It is an example of a country where an initial strong interest in oil distracted the government's attention from the need to treat gas on its own merits. Later, government interest became aroused in making better use of gas, and more structured policies gradually emerged; in particular since 1983 efforts have been made to use associated gas, in order to avoid flaring. But to begin with policies failed to adapt quickly enough to changing circumstances. Ultimately, in 1986–7 a serious attempt was made to design a 'Model Gas Contract', which defines the terms of contracts for the exploration and development of gas resources by companies, and for the sale of gas to the Egyptian national oil company (EGPC). In parallel with these problems which delayed gas exploitation there has been a permanent state of latent conflict between the government and aid donors, which has resulted in problems in financing infrastructure.

Through the 1960s and 1970s the government's energies were directed towards the oil market. Egypt had been one of the earliest countries to discover oil, but found itself losing out to newcomers on the oil scene. Policies were designed to try and match this challenge; even socialist President Nasser opened the doors to multinational oil companies, in an attempt to develop the country's oil resources. In this environment very little thought was given to natural gas. Those developments that did occur in the 1960s and 1970s, for example at Abu Madi, were more or less opportunistic and conformed to no overall energy plan. As the energy scene began to change in the late 1970s government awareness of gas potential grew. In late 1979 one company expressed an interest in LNG export possibilities, and by the early 1980s the government became interested in substituting gas for oil in the domestic market, thus releasing more oil for export to earn hard currency. However ideas for domestic utilization were still rather vague and the volumes used were small relative to the resource base. In 1982 known gas reserves represented 40 per cent of known oil reserves, but production was only 9 per cent of oil production. Policies have gradually become more defined; but there is still a credibility gap, created by the divergence between stated government priorities and what actually happens. For example, political mileage has been gained from the expansion of the residential market for natural gas. In practical terms this market sector only accounts for a fraction of consumption, although substitution with natural gas does play a useful role by reducing the need for LPG imports. The total volume of gas used in the residential sector is nevertheless extremely unlikely to overtake the bulk consumers such as power stations and industry – which the government is in fact heavily promoting, but with less publicity.

Raising finance for gas projects has also had its problems, and has often added considerably to delays that first occurred due to government vacillation. Decision-taking by many aid agencies is complicated and time-consuming in itself. There has also been a recurrent problem in reconciling the views of the agencies and the government on issues ranging from the methods of financing individual projects through to overall government pricing policy. Apparent political interference from outside agencies has often been a stumbling block to progress. The major difficulty in this respect has been the national policy of keeping consumer prices extremely low. Aid agencies are concerned about the distortion of relative prices in the economy, which encourages investment in uneconomic ventures, and in the long term weakens the economy as a whole. Loans have been refused or frozen, until such time as the government changes its policy.

Oil companies find themselves in a slightly different position. EGPC is now offering them a gas price related to international fuel oil prices, although the gas will be sold at a fraction of these levels to consumers. The oil companies are unhappy with this, lest future administrations find themselves unable or unwilling to continue with the subsidy, and therefore perhaps unable to pay the contracted price for gas. All other things being equal they would prefer consumer prices to be increased. But unlike the aid agencies they are not sufficiently perturbed to withdraw from the country. Reassured by a long history of being treated well by Egyptian governments they are generally willing to sign new gas agreements with EGPC.

Receiving payment is however only the first step for the companies, since they also wish to repatriate their earnings. This too presents a problem in Egypt, where there is a long history of strict exchange controls. The government circumvents this difficulty by making payment for gas in crude oil, rather than in cash – providing the company concerned is also an oil producer.

Important progress has been made recently. The new Model Contract drawn up in 1987 and awaiting ratification by the National Assembly is likely to encourage exploration for new reserves and the development of existing finds. Furthermore, a fairly comprehensive distribution network has now been constructed in Egypt, and gas is taking over an increasingly large share of the domestic energy balance.

There are lessons to be learned from Egypt about the need for coherent and responsive government policies, and also for the conduct of business between government, companies and funding agencies.

7.2 The Energy Balance in Egypt

Egypt has significant energy resources in the form of oil (3 billion barrels of remaining reserves), gas (9 tcf) and hydroelectricity.[1]

Hydroelectricity makes a significant contribution to the overall energy balance, representing 32 per cent of installed electricity generating capacity in 1985.[2] But its potential for further growth is very limited. Oil, meanwhile, has been a mainstay of the Egyptian economy, being both the major energy source used within the country and a valuable export commodity. In 1984/5 petroleum exports provided 79.3 per cent of the total value of Egypt's merchandise exports, at $4,955 million.[3] The importance of maintaining these export earnings has provided the impetus for greater concentration on gas development. For there is a danger that decreasing quantities of oil will be available for export. On the one hand oil production is beginning to decline, as the country's relatively small reserves are depleted, and on the other hand domestic energy consumption is growing very rapidly. This growth in consumption is partly due to population growth and to increasing industrialization of the economy, but can also be partly attributed to the high subsidies which have been given to energy in the past. These subsidies have allowed many industries to become established that would not otherwise have been economic, they have distorted the transport sector, and have discouraged conservation measures.

Consequently government attention has been increasingly focused on the country's least developed energy resource, natural gas. Production has grown rapidly since its first use in 1975, and it is planned to

[1] *BP Statistical Review of World Energy*, 1987; Ministry of Petroleum, Egypt.

[2] USAID, direct communication.

[3] World Bank, internal paper.

continue to increase the share of gas in the internal energy market, in order to preserve oil for export markets.

7.3 Gas Reserves, Production and Infrastructure

(a) *Reserves and Production.* EGPC put known gas reserves in 1986 at 9.36 tcf, of which 6.56 are non-associated and 2.8 associated. The government hopes to boost this figure to as high as 15 tcf by 1990, but this is thought by some oil companies to be over-optimistic.

The gas is found in four main locations. Two of the main gas fields are located in the Delta region – Abu Qir close to Alexandria, and Abu Madi further east towards Damietta. Abu Madi was the first major gas field to be discovered, in 1967, by Petrobel. This company is now extending its exploration north-eastwards off shore, where the Tina field is to be found. Other offshore fields in this area include El Temsah, Alif and Naf. The second main area is the Western Desert, where there is a major field at Abu Gharadiq, as well as the smaller Badr el-Din. Thirdly, the Gulf of Suez has considerable deposits of associated gas, which are collected to Ras Shukeir; and fourthly, Belayim on the Sinai side of the Gulf also has associated gas. Gas production has risen fast over the past twelve years, as shown in Table 7.1.

Table 7.1: Gas Production in Egypt. 1975–86. Billion Cubic Feet.

Year	Production
1975	1.6
1980	75.4
1981	84.4
1982	94.1
1983	107.6
1984	136.6
1985	172.1
1986	199.4

Source: EGPC, direct communication.

Most of the fields contain wet gas. The condensates and LPG are of great importance to the Egyptian economy, given the current need to import large quantities of LPG. Production of gas, condensate and LPG from the various fields in 1986 is shown in Table 7.2.

The new five-year plan, announced in January 1987, aims to increase gas production by 76.2 per cent, from 735 mcf/d in 1987 to reach 1,295 mcf/d by 1992. There will also be a tripling of LPG

Table 7.2: Egyptian Production of Gas, Condensate and LPG by Field. 1986. Thousand Tonnes.

	Natural Gas	Condensate	LPG
Western Desert	843	198	88
Abu Madi	1,655	212	–
Abu Qir	1,211	164	27
Shukeir	564	74	95
Sinai	33	15	21
	4,306	663	231

Source: *Al Bitrul*, Feburary 1987, p. 41.

production. (Meanwhile crude oil production is forecast to decline slightly, from 44.1 million t/yr to 43 million.)[4]

(*b*) *Infrastructure*. Egypt is fortunate in that an influx of dollars after 1973 allowed it to construct a good network of gas pipelines. While there are still important gaps, and the system has grown somewhat haphazardly, there is nevertheless a good basis for a comprehensive gas distribution grid. The government is currently working towards completion of the grid. This is important, since it will allow full flexibility of supplies from the various fields to the markets and load sharing. In 1975 there existed less than 30 miles of pipeline, but by 1982 this had expanded to 250 miles, and reached nearly 1,250 miles by 1987. Construction is still continuing, and a further 1,062 miles will be required to complete the national grid.

The first lines to be built were those linking gas fields to their local markets (luckily these distances were not very great), namely:

(a) Abu Madi to Cairo and the industrial complex at Helwan (and the power station just north of Cairo at Shoubra el Kheima). This line also reaches west, nearly as far as Abu Qir.
(b) Abu Qir to Alexandria.
(c) Abu Gharadiq to Cairo.

Additional lines have been built as follows:

(d) Ring east and south of Cairo, which will soon go right around the city to join in with the Abu Madi line.
(e) A gas-gathering system in the Gulf of Suez, taking associated gas from the Shukeir area to the city of Suez and the Ismailia area.
(f) From Ismailia to join the Cairo ring.

While the details of the pipeline network are important to the oil companies, to enable them to plan production and assess their access to markets, such information is not always easily available. This is a hindrance to gas development, and clarification of the infrastructural facilities will be important to any new gas agreements with the companies.

7.4 The Government Sector

Egypt is similar to other countries in that the centrality of energy policy to all other economic policies means that the government's whole approach needs to be assessed, rather than simply considering

---

[4] *World Gas Report*, 27 February 1987.

the energy sector in isolation. For example, a general problem with the Egyptian government sector is that there is a lack of co-ordinated planning between the various interest groups. While channels do exist to bring together the work of the various ministries these often seem to be blocked. For example, energy plans cannot be successfully drawn up in isolation from industrial planning, since these two sectors are so highly interdependent. The aluminium plant at Nag Hamadi in Upper Egypt consumes enormous amounts of energy; but consideration of its future falls outside the remit of the Ministry of Petroleum. Similarly, an overall policy of price subsidies to benefit the poorer sections of the community has led to distortions in the energy sector, which in the past has directly impeded gas development through limiting the potential returns to be earned by oil companies.

Companies have also encountered difficulties in penetrating the government bureaucracy. There are many layers of rules and regulations, and identifying the right government channel and procedure to follow is a difficult matter. The government has attempted to improve this; and it is true that companies which have been involved in Egypt for many years have learned their way around the system. Newcomers however may find it a real deterrent to becoming involved in the country, for they will need to work with the government from the very early stages of a project, to get import licences for equipment, work permits for staff, and so on, right through to checking their accounts and tax liabilities. They may find they can use their Egyptian joint-venture partner to sort out problems. But there is a definite need for procedures and practices to be simplified.

(*a*) *Ministry of Petroleum.* The Ministry of Petroleum was established in 1973, to oversee the overall management of the oil and gas sector. It provides a link between the government and the national oil company (EGPC), whose work it supervises. There is a separate Ministry of Electricity, which is wholly responsible for the power sector. EGPC does have considerable autonomy vested in it; for example it can establish its own internal regulations for financial, administrative and technical management, and can set its own employment and remuneration rules. But it is still subject to ministerial control in the final analysis. All resolutions of EGPC's Board must be approved by the Minister; and the Minister can, at his discretion, amend or cancel any Board resolution.

For three or four years before mid-1987 this separation of powers was limited however, since the Minister simultaneously held the position of Chairman of EGPC. This had important consequences, because the healthy interaction between policy-making and executive bodies was lost, and the role of the Ministry as a controlling body over EGPC was severely weakened. In addition it tended to stultify decision-taking within EGPC, as deference replaced independence. Decisions were also delayed even more than they would otherwise have been simply because the Minister had extensive responsibilities. This situation no longer obtains however.

(*b*) *Egyptian General Petroleum Corporation (EGPC).* Egypt's gas development has been almost entirely managed and carried out by EGPC. It functions essentially as a holding company, and oversees the entire spectrum of the oil and gas industries, through its various affiliates and joint ventures with foreign oil companies. It derives its funds from its share in the profits of its affiliates and joint ventures, and from supervisory and administrative fees. It has also raised money through government loans; and as a distinct corporate entity can borrow on its own account from external financial agencies.

EGPC has responsibility for all the following areas:

(a) general policies on granting concessions, and negotiating and preparing concession agreements (for approval by the Ministry);
(b) supervising exploration and exploitation by the oil companies;
(c) co-ordinating production,refining, transportation and distribution;
(d) exporting/importing as necessary;
(e) pricing of petroleum products;
(f) checking the accounts of oil companies.

While major decisions must be approved by the Ministry, EGPC nevertheless carries considerable power.

A specialized gas unit was formed within EGPC in 1975, which was upgraded to a department in 1982, headed by a Vice-Chairman for Natural Gas. This department now has responsibility for all technical and economic studies, planning and project development, review of cost estimates and approval of all plans for construction.

There are also several subsidiaries of EGPC concerned with gas. The *Petroleum Pipeline Company* supervises construction of the pipelines, and also arranges the distribution and sale of gas to large consumers in the industrial and power sectors. *Petrojet*, another subsidiary of EGPC carries out the actual construction. *Petrogas* is responsible for connecting households to the gas network, and distributing gas and LPG to the residential sector.

7.5 The Private Sector

All geophysical surveys and exploration are carried out under production-sharing agreements, or through joint ventures, with EGPC. Once a commercial oil or gas discovery is made EGPC sets up a non-profit-making operating company in association with the oil company concerned, and this company then operates and works the concession.

There is a very large number of foreign oil companies exploring and operating in Egypt under such arrangements with EGPC. The largest interest is held by Amoco, which through its joint-venture company, Gupco, produces 72 per cent of Egyptian oil. AGIP, through Petrobel, holds second place in oil production, with about one-third Amoco's quantity. Others include Arco, BP, Elf, Marathon, Mobil, Phillips and Shell. Those companies most directly involved with gas policy, as a result of their non-associated fields are AGIP (Abu Madi), Amoco (Abu Gharadiq), Phillips (Abu Qir) and Shell (Badr el-Din). The nature of Egyptian gas policy however, and the frequency with which associated gas has been found, have meant that all the companies have been acutely aware of gas policy in Egypt.

During the 1970s there were no specific contract terms applying to gas. In theory gas was therefore treated in exactly the same way as oil, although in practice arrangements were flexible. Under production-sharing terms for oil with EGPC cost recovery was allowed up to 40 per cent of profits, and the company took between 9 and 15 per cent of the remainder (depending on the exact terms of each concession). At this time however AGIP was producing gas from its Abu Madi field, and essentially giving it away to EGPC. AGIP added the condensate production from the field (5,000 b/d) to its oil production from other fields, and EGPC acquired the gas at no cost. Meanwhile Amoco was producing 110 mcf/d of gas from Abu Gharadiq, and selling it to EGPC. At the same time Phillips still produced gas from Abu Qir according to arrangements which pre-dated production sharing. While this informal system worked satisfactorily for some time, changes in external circumstances led to both sides seeking clarification of their rights.

7.6 Gas Policy, Contracts and Pricing

(*a*) *The National Reserve.* In the late 1970s some companies recognized a possible export outlet for Egyptian natural gas in southern Europe. The price of oil was high, with possible shortages envisaged. At the same time negotiations between the Algerians and others over LNG supplies were proving complicated. AGIP, especially, became interested in the idea of exporting gas from its Egyptian fields to Italy, and sought a legal framework from the government which would facilitate this. At the same time the Egyptian government was becoming more aware of its long-term interest in using natural gas within the country, in order to extend the life of oil exports. The prospect of gas exports therefore caused some alarm.

In 1980 EGPC first enunciated its new approach to gas exploitation. The Director General announced that Egypt was committed to establishing a National Reserve of 12 tcf of gas destined for use on the home market. No exports would be allowed until this target had been reached, but thereafter companies would be free to sell their gas as they wished and the normal production-sharing terms would apply. (Reserves were at that time about 4 tcf.) In addition EGPC assumed ownership of the first 12 tcf of gas, without compensation to the oil companies.

This policy was fraught with difficulties. The concept of the reserve was only vaguely defined: it was unclear whether small, scattered reserves of associated gas would be counted towards the 12 tcf, or only those fields that could be economically exploited. Moreover there was no discussion of how the gas might be used within Egypt – which market sectors would be addressed, or who would carry out the production and distribution. The company which actually discovered the gas had no right to develop the field, even if it saw a local market opportunity, since it was forced to surrender the gas to EGPC. But EGPC lacked the capability of producing the gas without assistance from an oil company. Of even greater impact was the fact that foreign oil companies would be forced to surrender gas fields without any compensation in return. This meant that from the companies' point of view discovering gas was exactly equivalent to a dry hole; this was a great disincentive to oil exploration in gas-prone areas.

In response to the companies' dissatisfaction, the government introduced a new compensation clause in 1982. Under this scheme if a company found at least 245 bcf of gas in an exploration area, and no oil, then it would be reimbursed its exploration costs plus interest, over a period of four years. The payments could be either in kind (i.e. oil) or in foreign exchange. While such compensation was obviously an improvement on nothing, it still did not arouse the enthusiasm of oil companies. At most, it would cover their costs, where they were actually looking for profits. It also seemed doubtful whether EGPC would actually be in a position to pay compensation in cash, given the large sums that could be involved, and the fact that such payments would be clearly visible in the accounts, perhaps causing political difficulties. Compensation in oil is much less visible, and is normally quite acceptable to companies. Both Elf and BP have taken compensation for gas finds in this way. Nevertheless, companies do not see their business in terms of just covering costs; consequently this compensation scheme did not encourage a high level of activity. The compensation scheme was also decried within Egypt, as a system open to abuse. For example critics suggested that oil companies had been compensated for more than they actually spent. Clearly the government wished to protect itself from such allegations. Moreover, given its falling oil reserves (about twelve years' remaining production in 1986) EGPC is anxious that extensive exploration should continue.

(*b*) *New Model Exploration Agreement.* In response EGPC drew up a model contract, with explicit terms on gas discovery and exploitation. The target national reserve of 12 tcf was also abandoned. EGPC began to negotiate the model contract with Amoco in 1986, with input also from other interested companies. The companies were enthusiastic about the concept of such an agreement, but working out the details proved a lengthy process. The first contract to be signed under the new terms was with Shell, in March 1987, to cover their Badr el-Din concession.

The new agreement offers production-sharing terms for gas and LPG similar to those for oil (i.e. 75/25 in favour of EGPC). This reintroduces the possibility of making a commercial return on gas finds. The periods allowed for exploration, relinquishment and development of gas fields have been extended. Most importantly, it commits EGPC to do its utmost to find local markets for the gas, and to take or pay 75 per cent of the quantity of gas stipulated in the sales contract. The price paid for gas will be determined in the following manner. The volume of gas or LPG will be converted to fuel oil in terms of British thermal units and the price for gas will then be 85 per cent of international prices for average sulphur fuel oil (95 per cent in the case of LPG). The 15 per cent discount is intended to cover EGPC's expenditure on providing the gas pipeline network. Payment may be either in cash or in crude oil.[5]

(*c*) *Pricing.* A major difficulty in promoting gas use in Egypt has arisen from the government's consumer pricing policy. EGPC is now offering companies a well-head price related to the international fuel oil price. While this in itself is welcomed (despite quibbles over the size of

[5] *Middle East Economic Survey*, 30:45, 17 August 1987.

EGPC's discount for transportation of the gas, etc.), the companies are still wary. For consumer prices charged by EGPC are considerably lower than such an internationally referenced price. The difference could be a price of about $3 paid to the oil companies, compared to $0.20 charged to customers. Despite the government's good record in keeping to its promises some doubt must remain in the companies' minds over the long-term reliability of such subsidies, however much they may trust those in power when agreements are negotiated. There will always be a risk of default on payment by EGPC, unless consumer prices can be brought nearer to the price paid to the producers.

The problem is that these subsidies extend over a very wide range of goods in Egypt, and most particularly to the major competitive energy source, oil. In 1983 the domestic price of oil was just one-fifth of the international market price. Despite considerable pressure from foreign interests, such as the World Bank, consumer energy prices were still only about 33 per cent of world prices in 1986. Government subsidies in that year were costing about $4 billion.[6] Apart from this cost to the government, serious distortions to the economy have resulted from the pricing policy. A number of industries have become established that would be quite uneconomic if charged the true costs for their energy use. The most glaring example of this is the aluminium plant in Upper Egypt.

However there seems to be more or less a stalemate over price reform in Egypt. Foreign companies are in theory anxious to see their well-head price ensured, through realistic consumer prices rather than government subsidies, but in practice continue to do business with EGPC. Meanwhile the IMF and the World Bank are concerned to reform the economy according to market principles, and have frozen all new loans. But the Egyptian government perceives insuperable political problems in altering fuel prices by the amount that would be required to reach international levels, and also sees internal pricing policy as its private concern. Prices have in fact been gradually modified, but it is also true that the government is still extending its policy of subsidizing consumers – for example, through pricing gas for residential consumers at extremely low levels (see below). Such actions cannot help in reassuring the companies about the likelihood of significant price reforms in the foreseeable future.

7.7 Markets

Gas consumption in 1985 amounted to 172 bcf. This was distributed

---

[6] World Bank, internal paper.

over economic sectors as shown in Table 7.3. This follows the pattern typical of many developing countries, with bulk users (and especially the power sector) providing the justification for expensive infra- structure. Government policy on priority sectors for gas consumption has however been very confused. Official statements have often been at variance with what has actually happened, and have frequently re- flected political rather than economic priorities. For example, the government ranks its priority uses for gas as follows: residential in first place, followed by fertilizers, then iron and steel, with power close to the bottom of the list. There has also been a consistent policy to encourage industrialization of the country, through heavy subsidies. Nevertheless, as Table 7.3 shows, most gas has in fact gone to those sectors where it is usually most economically beneficial.

Table 7.3: Egyptian Gas Consumption by Sector. 1985. Million Cubic Feet.

	Volume	*%*
Power	99,765	58
Fertilizer	41,672	24
Cement	14,656	8
Industry	11,831	7
Refineries	2,755	2
Residential	1,377	1
	172,056	100

Source: EGPC, Annual Report, 1985.

(a) *Power.* At the end of 1985 32 per cent of installed generating capacity was in hydroelectric plants, with 49 per cent in steam tur- bines and 19 per cent in gas turbines.[7] Gas provided 38 per cent of thermally generated electricity (compared to 54 per cent from fuel oil, 7 per cent from solar energy and 1 per cent from naphtha). The major gas fields all supply power stations already, including the largest gas-fired power station in the Middle East, at Shoubra el Kheima near Cairo. This was initially built with three 315 MW units, but a fourth unit is now being added. A number of other new plans are under consideration: a 1,200 MW plant near Damietta, the use of gas for peak loads at Nasr City and Heliopolis, and the piping of gas from Salam in the Western Desert to the towns 50 miles to the north.

Gas has thus made considerable inroads already into the power

[7] Most information in this section was obtained from USAID.

sector. Future growth of the sector will come from two sources. First there is room for further substitution for fuel oil which could otherwise be exported (3.6 million tons were burned in power stations in 1985). Secondly, the system is already operating close to capacity, and demand is expected to continue to grow very rapidly. New plant is constantly being added, and gas could take up a large part of this new demand. Between 1974 and 1984 total electricity generation grew by 12.4 per cent per annum, but in December 1984 the electricity authority was still compelled to shed loads of 600 MW during peak periods.[8] Load shedding is no longer a problem, due to the addition of new capacity, but there is no room for complacency. Demand will continue to grow fast, thanks to the very low pricing policies of the government, as well as general growth in the population and in the economy.

Thus there remains much room for further penetration of gas in this sector, although there is debate whether imported coal would be an equally good substitute for fuel oil, reserving gas for 'higher-value' uses. While coal is available cheaply on international markets its use in Egypt would involve building a whole new infrastucture, including specialized ports. The argument over gas versus coal hinges on the nature of the industrial ('higher-value') market, which is strongly criticized by some (see below). The only other indigenous source of energy is hydro, but remaining hydro potential is very limited. All that of the Nile has already been harnessed (about 80 per cent of the total potential) and plans for other schemes in the Qattara Depression have been shelved due to environmental worries. The nuclear option has been discussed, but is of course constrained both by cost and by international politics.

(*b*) *Industry.* Gas consumption in industry is dominated by a few very large consumers, notably three metals factories, two textiles manufacturers, and the construction industry. Over the last decade the latter has been a particularly flourishing sector of the economy, thanks to the influx of wealth from oil revenues and emigrants' repatriated earnings. The cement industry has grown very rapidly, and has substituted gas for fuel oil to a very large extent. The aim is to connect more large industrial consumers to the pipeline network over the next few years.

Consumption in the industrial sector is one cause of disagreements between the government and aid agencies. For example the government has long talked of converting the aluminium plant at Nag Hamadi to burn gas. Outside agencies see the plant as an economic

---

[8] World Bank, internal paper.

white elephant under any circumstances – its expensive products being traded for poor-quality East European tractors and so on. To burn valuable gas in the plant would be to add insult to injury in their view. The consequent disagreements partly explain the divergence between official government policy of great expansion in industrial gas use, and the actual history of expansion in the power sector.

(*c*) *Fertilizers.* The very first use of natural gas in 1975 was in a fertilizer factory, and there are now three factories producing nitrogenous fertilizers. Two plants at Talkha, using gas from Abu Madi, have a total capacity of 550,000 t/yr of ammonia; and a plant at Suez using local associated gas has a capacity of 400,000 t/yr of nitrogen. This production is destined entirely for the home market, since there is very high demand for fertilizer in Egypt. Although the quantity of arable land is limited, it is very intensively cultivated with widespread irrigation, and consequent need for nitrogenous additives. Consumption of fertilizer in Egypt in 1984 was 364 kg per hectare of arable land, compared to an average of 123 kg/he in industrialized market economies and 56 kg/he in middle-income developing countries as a whole.[9] Given the availability of gas as a feedstock the policy of becoming self-sufficient in nitrogenous fertilizer makes good economic sense.

(*d*) *Residential.* Unusually for developing countries the government has heavily promoted the distribution of gas to households, in an effort to reduce the level of LPG imports. (In the early 1980s 70 per cent of all LPG was imported.[10]) Supplies have been taken to four outer suburbs of Cairo: Helwan and Maadi to the south-east of the city, Heliopolis to the north-east, and Dokki to the west. These are areas with new high-rise developments, housing middle-income sectors of the population. Approximately 300,000 households are now connected.[11]

The system was constructed by British Gas for Petrogas; but there has been considerable argument over its design. British Gas, and other outside bodies such as the World Bank, advised against supplying households, and in favour of concentrating on the commercial sector (hotels and offices) where demand per customer would be higher. For most households will use gas solely for cooking, consuming a minimal amount of gas in comparison to the infrastructural costs of connection; whereas hotels and richer households can also use gas

[9] World Bank, *World Development Report*, 1987, p. 212.
[10] *World Gas Report*, 30 June 1986.
[11] *Middle East Economic Survey*, 30:35, 8 June 1987.

for water heating, increasing considerably the consumption per connection. However the government felt that the political advantage to be gained from supplying a large number of people with gas outweighed the economic disadvantage.

In order to attract custom amongst the middle classes however, it was necessary to price the gas at or below the competing fuel, LPG, which is itself heavily subsidized. Petrogas carried out a survey, which showed that the average household used the equivalent of 1.5 bottles of LPG a month for cooking purposes, for which they paid about E£0.30. The first tranche of gas consumption is therefore priced at this level. Prices for additional quantities are sharply progressive. Most of the households which have been connected have not yet had meters installed, presumably because of the high cost involved in installation compared to the expected takings. This is effectively an admission of defeat, in that no attempt can then be made to collect payment from customers. It is arguable that a flat charge would be more appropriate, as used in Pakistan. While some wastage may then occur, through appliances being left turned on, the same is true of the current situation, where no money at all can be collected.

(*e*) *Transport*. Little interest has been shown in introducing a programme of CNG substitution for transport, although a small pilot project has been run.

(*f*) *Exports to Israel*. The possibility was mooted in 1986 of a 325-mile pipeline to take gas (currently flared) from the Gulf of Suez to industries at Beersheba and the power plant at Ashdod. The idea has not been taken up with any enthusiasm by the government, which sees it more as a political proposal than a viable economic project.

7.8 Finance

Much of the infrastructure in Egypt has been funded through USAID and multilateral organizations such as the World Bank. While the World Bank has often baulked at providing assistance under current pricing policies, nevertheless it has lent substantial sums to the country. For example it lent $75 million towards the gas-gathering system in the Gulf of Suez (EGPC covered the remaining $95 million from its own resources). Similarly the IDA contributed $50 million towards the residential distribution system. Both the World Bank and the European Investment Bank contributed towards the $165 million costs of the second-stage development of the Abu Qir gas field.

Egypt's financial position causes severe problems for oil companies, especially if they are considering developing gas for internal markets, as opposed to exporting crude oil. Egypt is a very supply-constrained country, especially for capital. This gives rise to difficulties in implementing projects – importing materials becomes problematic, bottlenecks are encountered in the provision of infrastructure, and local auxiliary services tend to lack vital components. This can be very frustrating.

Even worse, the repatriation of profits is threatened. Egypt has a history of very strict exchange controls; these have recently been liberalized, but the future does not look reassuring. The foreign debt is very large, probably in excess of $40 billion. In the past this debt consisted mostly of long-term government liabilities; but over the last two years the debt structure has worsened. There is now a much higher proportion of short-term commercial debt, which could well lead to rescheduling problems. Furthermore the balance of payments is suffering, both from a reduction in income from oil exports, and from smaller remittances from nationals working elsewhere in the Middle East. Oil companies are therefore very worried about the convertibility of any earnings they may make on gas in Egypt. There is a partial solution to this problem, providing a company produces oil as well as gas. This is to make payment for the gas in the form of additional oil allowed to the company. Such a policy has certain administrative advantages, but it should not be forgotten that payment in a highly tradable good (oil) is equivalent to payment in foreign exchange and does not therefore alleviate the foreign exchange constraint which affects the economy. EGPC has shown itself willing to follow such a flexible course. But there must be a limit to the amount of oil that the government can use in this way. Should gas production really fulfil its potential, it is questionable whether such a method would in itself be sufficient to alleviate the companies' concern.

Sources

In addition to the sources quoted in the footnotes, much of the information in this chapter was obtained from unpublished World Bank material, and from interviews with executives from the following companies and organizations: AGIP, Amoco, British Gas, BP, EGPC, Elf-Aquitaine, Marathon, Shell, USAID, and the World Bank.

APPENDIX TO CHAPTER 7

Background Data

Population[a]: 48.5 million
GNP per capita[a]: $610
Average annual growth in GNP per capita, 1965–85[a]: 3.1 per cent
GDP[a]: $30,550 million

External public debt[a]: $18,501 million
External public debt as percentage of GNP[a]: 64.5

Total energy consumption[b]: 15.726 mtoe
   Commercial: 92 per cent   Non-commercial: 8 per cent
Commercial energy consumption per capita[a]: 588 kgoe
Commercial energy imports[c]: 1.686 mtoe
Commercial energy exports[c]: 8.209 mtoe

Proven Reserves and Production[d]

	Gas	Oil	Coal	Hydro
Proven Reserves	9.4 tcf	3.6 bbls	–	2,660 MW
Production	199 bcf	820,000 b/d	–	10,400 GWh

Consumption by Sector: Thousand Tons of Oil Equivalent[c]

	Gas	Oil Products	Coal	Total	% Share
Power	–	2,156	–	2,156	14
Industry	1,347	5,553	483	7,383	47
Transport	–	3,375	–	3,375	22
Households	5	1,805	–	1,810	12
Other	–	862	–	862	5
Total	1,352	13,751	483	15,586	
% Share	9	88	3		100

Power Sector[f]

Installed capacity: 8,331 MW
Electricity generation (%): Hydro=32, Oil=37, Coal=0, Gas=26

Sources

(a) World Bank, *World Development Report*, 1987, pp. 202–237. Figures refer to 1985.

(b) OECD, *Energy Balances of Developing Countries 1971/1982*, 1984.

(c) EGPC, *Annual Report*, 1985.

(d) Gas reserves and production, Ministry of Petroleum, Egypt. Oil reserves and production, *BP Statistical Review of World Energy*, 1987. Hydro reserves, World Bank, *The Energy Transition in Developing Countries*, 1983. Figures for gas and oil refer to 1986, and hydro to 1982.

(e) OECD, *Energy Balances of Developing Countries 1971/1982*, 1984. Figures refer to 1982. Consumption of gas in 1985 amounted to 3,692,000 toe, of which 2,130,000 toe was consumed in the power sector, 1,531,000 toe in the industrial sector, and 31,000 toe in households. Total consumption of oil products in 1985 amounted to 21,900,000 toe. (EGPC, *Annual Report*, 1985.)

(f) Figures supplied by USAID; refer to 1986.

8 MALAYSIA

8.1 Country Summary

The history of gas development in Malaysia throws into clear relief the problems that are encountered in implementing a gas utilization scheme, even in a relatively fortunate developing country. The reserves of natural gas in Malaysia are very large, and an LNG export scheme has been functioning successfully since 1983. The country's economy has been developing at a steady pace, and there are no problems in raising capital to finance gas infrastructure. But nevertheless internal use of the gas has grown only very slowly. The major pipeline network has still not been built, although it has been under consideration since 1982.

There is a good incentive to encourage gas use, in that the country's oil reserves are depleting fast. Over 60 per cent of current energy consumption relies on oil,[1] so a substitution programme could release significant quantities of additional oil for export, either now or in the future. The government is also anxious to insure against the impact of future oil 'shocks' by diversifying its energy sources.

The major energy-consuming sector in Malaysia, as elsewhere, is power generation, which currently relies heavily on fuel oil. Gas is a good candidate for substitution, since the large hydroelectric potential which does exist is located in Sarawak, and is thus inaccessible to the major markets. There is an ambitious plan to pipe gas throughout the Peninsula in order to supply the major power stations, and economic studies have supported this strategy. The government however is also keen to introduce cheap imported coal as a fuel for the power sector. This has led to a debate between the national oil company, which is both well entrenched in the system and powerful, and the government administration.

Other aspects of gas use in Malaysia are also of interest. First there is a plan to export gas to Singapore by pipeline. This will simultaneously bring gas to a relatively underdeveloped area of Malaysia which would otherwise be unlikely to merit a gas supply, and earn

---

[1] Most information in this section was obtained from Petronas.

hard currency through the sale of gas to a neighbouring country. Secondly, there are lessons to be learned from the problems encountered in Sabah in using associated gas. Here the state government has attempted to make use of gas which would otherwise have been flared, by building heavy industries for which all other raw materials needed to be imported. Predictably this has led to severe financial problems, as the industries have struggled to compete on international markets. There is now greater recognition of the limits on using associated gas, where markets are severely limited. Emphasis is instead being put more on limiting the production of oil, where undue amounts of associated gas would be wasted in the production process.

8.2 The Energy Balance in Malaysia

Malaysia has historically been very dependent on its oil reserves, both as a source of export earnings and as an energy resource within the country.

Table 8.1: Commercial Energy Consumption in Malaysia. 1980 and 1985. Percentage Shares.

	1980	*1985*
Petroleum Products	88	62
Natural Gas	–	17
Hydroelectricity	12	19
Coal	–	2

Sources: Petronas and World Bank internal document.

In the future however this consumption pattern will need to change radically, if Malaysia is to maintain its relative independence from imported energy. Malaysian oil reserves are depleting fast. There are now about 3 billion barrels of proved reserves left, which will last for about another seventeen years at the 1986 production level of 500,000 b/d. Prospects of finding further major reserves are not high. Consequently, by the end of the century it is likely that Malaysia will have become an oil importer (it already imports diesel due to a refinery imbalance).

There is considerable hydroelectric potential, at 29,000 MW, but unfortunately this is mainly located in Sarawak, while the population centres are in peninsular Malaysia. While it would be technically possible to transmit electricity under sea from Sarawak this would be an extremely costly enterprise.

It is therefore seen as important that the huge gas reserves of 53 tcf are developed, and that as little associated gas as possible is flared. Government policy is that 39 per cent of the country's energy needs should be supplied by gas by 1990. Recent delays in implementing the Peninsular project however probably mean that this target will not be reached by that date.

8.3 Gas Reserves, Production and Infrastructure

(a) *Reserves and Production.* Total proven gas reserves in Malaysia amount to nearly 50 tcf, placing it in fourteenth position in the world.[2] Malaysia thus has very large quantities of natural gas available to it.

It is important to note that the reserves are found in three distinct locations, only one of which is relatively close to the major population centres of the country. This area, the largest of non-associated gas lies offshore the peninsular state of Trengganu, which has total reserves of about 21 tcf. Central Luconia, the first large gas province to be discovered, lies offshore Sarawak; it contains 14.5 tcf of non-associated gas. The remainder is associated gas, and is mostly found in the Shell acreage offshore Sarawak and Sabah, and in the Esso and Carigali acreage offshore the Peninsula.

Table 8.2: Gas Production in Malaysia. 1983–6. Billion Cubic Feet.

	Total Production	*Volume Exported*
1983	131	53
1984	325	166
1985	437	212
1986	528	240

Source: Cedigaz, *Le Gaz Naturel dans le Monde*, 1984, 1985, 1986.

Commercial production has been growing fast, as shown in Table 8.2. This shows that approximately half of all the gas produced so far has been exported to Japan as LNG; these volumes have now reached their full contract level, and demand for LNG is not expected to increase in the near future. Production for the home market will however increase should the next phases of the Peninsular pipeline project come to fruition.

The non-associated gas in Malaysia has the important advantage that production costs are relatively low, by international standards.

[2] *BP Statistical Review of World Energy*, 1987.

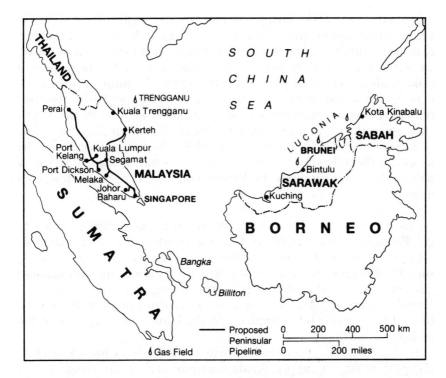

The fields are large, and are not too far off shore. A further important feature of the gas in Malaysia is that it is rich, containing up to 10–12 per cent condensates, and 8–10 per cent LPG.[3] This is important because LPG and condensates are much more easily tradable than gas. In particular, LPG is widely used in households and small industries, and has potential as a substitute for imported diesel in the transport sector. So long as sufficient supplies are available LPG requires much smaller expenditure on infrastructure than would the use of methane in these market sectors. The very first phase of gas development in peninsular Malaysia therefore included the construction of a gas separation plant at Kerteh, with a capacity of 250 mcf/d. Once the pipeline is extended in the next phase of the gas plan the capacity of this plant will be increased to 750 mcf/d.

The production of associated gas as a by-product of oil has called for special measures. As recently as 1984 as much as 95 bcf of gas were either flared or vented; this had grown from 65 bcf in 1983, and fell back to 75 bcf in 1986,[4] after implementation of the Sabah scheme

---

[3] World Bank, direct communication.

[4] Cedigaz, *op. cit.*, 1986.

described below. Most flaring now occurs in Sarawak, where there are very limited opportunities for using the gas. As a general policy Petronas requires contractors to reinject gas whenever this is possible. In addition oilfields are not allowed to be produced if the gas-to-oil ratio is above 3,000 scf per barrel. In Sabah, the joint-venture partners Petronas and Shell have come to an arrangement whereby they relinquish the gas they produce in association with oil to the Sabah state government, in return for a flat fee of $1 million plus costs. The quantities of gas involved are 90 mcf/d for the first ten years and 60 mcf/d for the following five years. This gas is now being put to use in Sabah, but unfortunately in loss-making enterprises as described below.

(*b*) *Infrastructure*. Malaysia's most ambitious plans for using gas are in the Peninsula, but they depend on the construction of an extensive pipeline network to transport the gas. This is being achieved in successive stages. First, a pipeline was built joining the two offshore gas fields of EPMI and Carigali, and taking their gas on shore to Kerteh, where a power station and gas processing plant have been built. This pipeline was completed in 1984, and has a capacity of 400 mcf/d.

In the next stage a 270-mile pipeline is to be built from Kerteh to the west coast, to supply Kuala Lumpur and several large power stations. This should be completed by 1988–9, and may later be extended to the north of the country.

The longest pipeline is planned to run through the southern half of the country to Singapore. While there are no large market outlets within the southern region of Malaysia the supply of gas can be economically justified on the basis of an export contract with Singapore. The plan is that Singapore should take 150 mcf/d for thirty years, or for as long as the gas lasts, to supply a 1,650 MW power station and perhaps also to be used as petrochemical feedstock.

The pipeline grid for the second phases will total over 470 miles, and together with a gas processing plant will cost $135 million. Petronas has no problem in raising this money, and studies by both Petronas and the World Bank have shown the scheme to be economically worth while at oil prices of $15–20 (although of course this will depend on the gas price eventually agreed with Singapore, as well as the pricing regime within Malaysia). There have been delays in implementing it however. There have been disagreements between the different provinces (each wanting to reap maximum advantage from the gas) and a debate over ownership of the pipelines. While Petronas has been anxious to maintain its monopoly, several private companies including Enron have expressed interest in constructing the line to Singapore.

The whole project was thrown into doubt however by the 1986 collapse in oil prices, which had a profound effect both on the economy of Malaysia, and on Petronas itself. In May 1986 it was announced that the project would be delayed by one year; but in September the government changed its mind and said it would go ahead as planned. By November Petronas was about to sign a contract with the Canadian firm Novacorp for management consultancy and operation and maintenance of the pipeline. The government suddenly stepped in however, and said the whole scheme was 'indefinitely postponed', which was widely interpreted as meaning that it was cancelled. Following pressure from Petronas the government withdrew this statement, and in December the contract was signed. It gives no indication of when work will start; but construction tenders are scheduled for a year's time, with completion by 1990.

As yet no sales contract has been signed with Singapore. Following some negotiating difficulties the two sides now seem to have agreed in principle however, following a meeting between the premiers of the two countries in the autumn of 1986. Nevertheless, details of pricing are yet to be determined. A shadow is also thrown over the scheme by continuing negotiations between Singapore's Prime Minister and the Indonesian President, over the import of gas from Indonesia's fields offshore Sumatra. Singapore favoured this source because the Indonesians put no restrictions on what they used the gas for, whereas Malaysia was insisting that its gas be used only in power plants and not for general commercial purposes. Petronas has now removed this restriction however, and it seems likely that Singapore will take gas from both Malaysia and Indonesia.

8.4 The Government Sector

The existence of a well-established and powerful national oil company in Malaysia creates room for energetic debate over energy policy, between the national oil company on the one hand and government officials on the other. Unlike some countries however the national oil company is enthusiastic over the potential for natural gas, and does not see it as a competitor for its traditional oil business. Petronas thus champions both oil and gas against government policies which sometimes favour other energy sources. For example the government is considering importing coal for the Kelang power station, while Petronas considers that gas should be used. Of course Petronas has a vested interest in encouraging oil and gas use over other fuels; but it also has a high reputation for efficiency and good management.

(*a*) *Ministry of Energy, Telecommunications and Posts.* This ministry was created in 1979 to co-ordinate the activities of the various government agencies dealing with energy. It includes the secretariat for the Cabinet Committee on Energy; this small group has the task of assessing overall supply and demand patterns for all types of energy, including oil. Its role is crucial to implementing a gas utilization policy, since it should have the capability to consider the problems caused by substituting gas for fuel oil, and similar issues. The secretariat services the *Cabinet Committee on Energy.* Formed in 1980 this Committee is responsible for evolving energy policy. It consists of the Prime Minister, the Deputy Prime Minister and the Ministers for Energy, Trade and Industry, and Science and Technology. It also has representatives from Petronas, the National Electricity Board, the Treasury and so on. By bringing together these various interests it has the potential to function as a very useful organ for the planned implementation of a natural gas plan.

(*b*) *Petronas.* Petronas was set up in 1974 under the Petroleum Development Act, to act as the national oil company with exclusive rights over the nation's oil and gas reserves, and as a means of transferring technology from the transnational oil companies to Malaysian nationals. It is now the country's biggest company, both in terms of earnings and tax payments. As such it has considerable power in relation to the government, although it has been noted for keeping out of politics as such, concentrating closely on its allotted business, oil and gas.

It has established a reputation of being well run and efficient albeit with limited expertise in the oil industry. Most of its senior staff came from the civil service and tend to be without political ambitions. Rather, they are said to have a somewhat conservative outlook, and to be resistant to outside pressures from multinational companies. Decisions tend to be slow, and Petronas is renowned for its negotiating skills and for striking hard bargains. However, once an agreement is reached they offer full co-operation with the companies. For example, on the Bintulu LNG project negotiations were said to be 'tough and arduous'; but once the deal was sealed there was close collaboration to get the project working on time. Similarly, production-sharing agreements were traditionally among the toughest in the world, and much more exacting than those offered by their neighbours in Indonesia, Thailand or the Philippines. For example, when Elf-Aquitaine signed its 1982 exploration agreement it was estimated that the company would make only $1.40 per barrel of oil, compared to an international average at that time of $2–4.[5]

The financial situation of Petronas is now rather poor, but this can be largely blamed on government actions. The government has used Petronas on a number of occasions to solve financial problems in other sectors of the economy. For example it rescued Bank Bumiputra from problems caused by its real estate speculations in Hong Kong. It also used Petronas to finance the purchase of aeroplanes for the state airline. These decisions came direct from the Prime Minister's office, and were reported to be very unpopular with Petronas management.[6]

Petronas Carigali was formed as an exploration subsidiary of Petronas in 1978, and given a 23,050-km^2 block offshore peninsular Malaysia, in an ex-Conoco acreage, where petroleum had already been found, which was both promising and low risk. As well as this concession area of its own Carigali has explored in partnership with BP offshore Sabah and with Elf offshore Sarawak. The company has thus developed local skills in exploration, and Malaysians are gradually taking over from the expatriates who originally managed it. Carigali discovered and operated the Duyong gas field off Trengganu, and is also responsible for the pipeline which transports gas from both Duyong and the EPMI field on shore to Kerteh.

8.5 The Private Sector

Under the 1974 Petroleum Development Act all rights to exploration and production of oil and gas are vested in Petronas. All foreign companies are therefore obliged to enter into production-sharing arrangements. Traditionally Petronas has bargained hard over the contract terms, but recently it has substantially improved the terms for new contracts, in an effort to retain the interest of hard-pressed companies.

Two main companies have been involved with gas in Malaysia, namely Shell and Esso. Shell drilled its first oil well in Sarawak in 1910, and has been producing oil in Malaysia for many years. It discovered the Central Luconia gas fields offshore Sarawak in 1970. It also has associated gas in its oilfields offshore Sarawak and Sabah. Esso is a much more recent entrant to Malaysia, having drilled its first well offshore Sabah in 1971. Its share of oil production has grown steadily however, until in 1984 the production quotas for the two

---

[5] *Financial Times*, 11 January 1983.
[6] *Platt's Oilgram News*, 26 November 1986.

companies were split 50/50. About half of the gas produced off Trengganu is from the Esso (EPMI) field, some of it associated gas that was previously flared, and some non-associated.

Very little actual exploration is now under way. Oil prospects are not thought to be very good, with the main fields already having been exploited. There is no immediate outlet, either within Malaysia or on the export market, for further gas discoveries. Furthermore prices for both oil and gas are low. The number of exploratory wells sunk fell from sixty-one in 1981, to twenty-one in 1983, twenty-seven in 1985, and fourteen for each year from 1986 to 1988.[7] Improved contract terms have helped to maintain some interest, despite the world-wide reduction in exploration effort: six groups of companies have signed Letters of Intent for new production-sharing agreements since their introduction.

8.6 Gas Policy, Contracts and Pricing

(a) *Contracts.* The original contract terms for Shell and Esso were fought out over two years, and were very stiff. The terms eventually settled on were that 20 per cent of oil and 25 per cent of gas production were to be allowed as cost recovery factors. Remaining production was split 70/30 between Petronas and the contractor. Royalties amounted to 5 per cent to the federal government and 5 per cent to the local state. Various bonuses were also payable.

These terms were gradually weakened over the years. In 1982 Elf won a concession in which cost recovery deductions were increased to 30 per cent for oil and 35 per cent for gas. By mid-1983 both Shell and Esso were looking for more acreage as their fields approached depletion, but they wanted better terms. It was not until the end of 1985 that such terms were finally offered. Cost recovery factors were increased to 50 per cent for oil and 60 per cent for gas. All bonus payments were abolished. Royalties stayed at 10 per cent, but the profit split was altered to give a better return to the contractor, especially on small reserves. For gas it was now put at 50/50 for the first 2 tcf to be produced and sold from a contract area, thereafter changing to the original 70/30.[8] However, these more generous terms were introduced just before the 1986 collapse in oil prices, and as a result they did not attract as much interest as they might otherwise have done.

---

[7] Petronas, direct communication.

[8] *Petroleum Economist*, January 1986.

(*b*) *Pricing*. An overall pricing policy for gas used within Malaysia is yet to be determined. It has been the subject of vigorous debate within the country, especially between the power authority (seeking advantageous terms) and Petronas.

8.7 Markets

The market for natural gas in Malaysia needs to be considered as three separate entities, given the distribution of the gas fields. Geographical location prevents the sharing of resources, and in any case the fact that gas is located adjacent to each of the three main areas of the country means that this is not in fact essential. The second important feature determining market development in Malaysia is the population distribution, which is heavily concentrated on the Peninsula.

The gas off Sarawak was the first to be developed, for an LNG export project. This was a self-contained scheme which succeeded thanks to market conditions at the time. There has been some local industrial use for many years, but further opportunities are very limited. It would require construction of new industries to use the gas, rather than being a fuel substitution exercise. In the second area, Sabah, the associated gas used to be flared; but there are now efforts to use some of this locally in both power and industry. Thirdly, the gas offshore peninsular Malaysia is destined for possible export to Singapore by pipeline as well as for extensive use in the power sector and industries within Malaysia.

(*a*) *Power*. The power sector in Malaysia has historically been supplied almost entirely by fuel oil, although the shares of both hydro and gas have recently increased. By 1985 oil accounted for 61 per cent of power generation, with hydro and gas taking 26 per cent and 13 per cent respectively.[9] Government policy has been to strongly encourage gas substitution in this sector. While hydroelectricity schemes have also been developed there is only limited potential for further growth in hydropower in those areas of Malaysia where the energy is needed.

The total hydropower potential is 29,000 MW, but 69 per cent of this lies in Sarawak and 17 per cent in Sabah. Only 14 per cent (4,060 MW) is in peninsular Malaysia, where some 83 per cent of the population is found.[10] Thus, while there is huge potential for hydro generation in Sarawak, demand in that state is relatively restricted.

[9] Petronas, direct communication.
[10] Asian Institute of Technology, unpublished paper.

There has been some discussion over the possibility of building an under-sea transmission line from Sarawak to the mainland, but this would be an extremely expensive project, and is unlikely to materialize. A large part of the potential in the Peninsula has already been tapped.

The introduction of gas to the power sector is following a two-stage programme. First, gas-fired power stations were built to provide the power for major new industries constructed under the government's industrialization policy. These included a 900 MW combined-cycle plant at Paka in Trengganu, which was built in tandem with a new steel plant, and a 70 MW plant in Sabah, using associated gas, to power a new sponge-iron smelter. As part of entirely new industrial schemes these did not directly substitute for fuel oil. The next phase of development, with the trans-peninsular pipeline, would however displace fuel oil in existing plants. Large power stations at Perai in the north, and at Port Dickson and Pasir Gudang would be converted from fuel oil to gas. A number of smaller power stations might also be converted, dependent on case-by-case evaluations of the economics.

The World Bank has estimated that by the year 2000 gas could provide as much as 81 per cent of Malaysia's power; with hydro providing 8 per cent, coal 10 per cent and oil just 1 per cent.[11]

(b) *Industry*. Gas is used in Trengganu, in the area surrounding the Kerteh separation plant, to fuel a steel mill and petroleum processing units (in addition to the Paka power station and some households). Once the pipeline reaches Kuala Lumpur the gas will presumably also be used there in local industries. However the major input of gas to the country's industries is envisaged in the form of electric power. It is in this sense that gas has been an integral part of the government's strategy of developing indigenous heavy industries, and avoiding over-reliance on its export commodities. The steel plant at Chukai falls into this category.

Some of these schemes have however been of marginal economic justification, and several have completely collapsed. For example the Sabah state government, against the advice of the federal government, built a gas-fired sponge-iron plant, which encountered considerable financial difficulties. This relied entirely on imported raw materials, with the exception of locally available natural gas. Another scheme, a $1 billion aluminium smelter for Sarawak, using imported alumina and local natural gas as fuel luckily never materialized. The methanol plant on Sabah has also been a commercial failure.

---

[11] World Bank, internal document.

(*c*) *Residential and Commercial.* Only 4 per cent of total commercial energy used in 1980 (492,000 toe) was consumed in this sector.[12] This comprised mostly kerosine, with some LPG. Only a very limited role is envisaged for natural gas in households: for example small amounts are used in the immediate vicinity of the separation plant at Kerteh. The production of natural gas is nevertheless important to this sector, since considerable quantities of LPG are extracted from the gas at the separation plant. In the absence of this production there would be a need to import LPG.

(*d*) *Transport.* The transport sector accounts for 46 per cent of all petroleum products consumed in Malaysia,[13] and demand is expected to grow steadily through to the end of the decade. Scope for gas substitution is limited however, due to the need for great care in balancing refinery output with demand. There is already a gasoline surplus in Malaysia, so there is little reason to encourage conversion of these vehicles to CNG. Diesel does need to be imported in order to meet demand; but the government is working primarily to convert diesel vehicles to LPG, rather than to CNG. Since the methane can be more profitably used in the power sector, and LPG is extracted from the natural gas, this policy makes considerable sense. A pilot CNG project has been carried out, with assistance from a New Zealand firm. Prospects in this market sector do not seem particularly large however.

(*e*) *Fertilizers.* An ASEAN ammonia/urea plant is in operation, but is reported to be running at a loss. Equity in this project is 60 per cent with Petronas, 13 per cent with each of the Indonesians, Thais and Filipinos, and 1 per cent with Singapore.[14]

(*f*) *LNG.* The LNG scheme is running successfully, but is effectively separated from any use of natural gas within the country. The scheme is run by MLNG, a joint-venture company between Petronas (60 per cent), Shell Gas (17.5 per cent), Mitsubishi Corporation (17.5 per cent) and the Sarawak state government (5 per cent). The plant was opened in January 1983, and is contracted to deliver 6 million tonnes of LNG per annum to Tokyo Electric and Tokyo Gas for twenty years. There are plans to expand the LNG project, but these are unlikely to be realized in the short term.

---

[12] Petronas, direct communication.

[13] *Ibid.*

[14] *The Nation*, Bangkok, 17 October 1986.

(g) Shell's Middle-distillate Plant. Shell has been negotiating with the Malaysian government to set up the first full-scale commercial plant using their middle-distillate synthesis process. This process has resulted from intensive research sparked off by the 1974 oil crisis. It converts natural gas first to syngas, and thence into such liquid transport fuels as naphtha for motor gasoline, kerosine for jet fuels or gas oil for diesel.

The Malaysian government has not been very enthusiastic at the prospect of being a technical pioneer, although the country does have a shortage of diesel, and despite the fact that Shell have offered to take more than 50 per cent of the equity. The plant has been costed at $820 million, and would use 100 mcf/d of natural gas. It would be sited at Bintulu.

Shell's most recently published estimates put the feedstock cost of the process at about $10 per barrel, and other fixed and variable operating costs at an additional $10. Given current oil prices, and given the general state of the Malaysian economy, with widespread cut-backs in the government's investment programme, there is some doubt over the future of the scheme, but negotiations between Shell and the government are still continuing.[15]

8.8 Finance

Malaysia has not had difficulty in attracting foreign investment. This has been especially true in the assembly-type electronics and textile industries, but has been felt too in other sectors of the economy. Investors have been attracted by perceived political stability, cheap labour, adequate infrastructure and fiscal incentives. This interest from foreign companies has been matched in the past by relative government prosperity stemming from commodity exports of oil, tin, rubber and palm oil. However, the government's determination to move away from the vulnerable role of a primary commodity exporter came just too late, and coincided with the collapse in prices of those commodities. Its policy of expanding industries such as iron and steel, cement, petrochemicals, machinery and car manufacture, all required substantial government investment, at a time when funds were suddenly short. The country's favourable debt/service ratio encouraged it to borrow large amounts in order to finance the industrialization programme, with the result that external debt increased threefold between 1980 and 1984.[16]

---

[15] Cedigaz, 10 January 1986.
[16] World Bank, internal document.

In reaction to this the government became very concerned at over-spending, and attempted to rein back public expenditure. The fall in oil prices in 1986 further exacerbated this trend. Investment in gas infrastructure is nevertheless seen as one way of restimulating the economy, and of conserving oil reserves for future exports. The government is thus faced by a considerable dilemma, which is reflected in the recent hesitations over phase two of the Peninsular pipeline project.

Sources

In addition to the sources quoted in the footnotes, much information in this chapter was obtained through interviews with executives in the following companies and organizations: Exxon, Petronas, Shell and the World Bank.

APPENDIX TO CHAPTER 8

Background Data

Population[a]: 15.6 million
GNP per capita[a]: $2,000
Average annual growth in GNP per capita, 1965–85[a]: 4.4 per cent
GDP[a]: $31,270 million

External public debt[a]: $17,966 million
External public debt as percentage of GNP[a]: 62.0

Total energy consumption[b]: 14.841 mtoe
   Commercial: 87 per cent   Non-commercial: 13 per cent
Commercial energy consumption per capita[a]: 826 kgoe
Commercial energy imports[b]: 6.731 mtoe
Commercial energy exports[b]: 24.945 mtoe

Proven Reserves and Production[b]

	Gas	Oil	Coal	Hydro
Proven Reserves	49.4 tcf	3.1 bbls	28 mt	29,000 MW
Production	526 bcf	484,460 b/d	–	5,800 GWh

Consumption by Sector: Thousand Tons of Oil Equivalent[b]

	Gas	Oil Products	Coal	Total	% Share
Power	497	2,331	–	2,828	24
Industry	1,982	2,102	360	4,444	37
Transport	–	4,219	–	4,219	35
Households	–	492	–	492	4
Total	2,479	9,144	360	11,983	
% Share	21	76	3		100

Power Sector[b]

Installed capacity: 4,822 MW
Electricity generation (%): Hydro=26, Oil=61, Coal=0, Gas=13

Sources

(a) World Bank, *World Development Report*, 1987, pp. 202–237. Figures refer to 1985.
(b) Petronas, direct communication. Figures refer to 1986, with the exception of power sector data, which refer to 1985.

9 NIGERIA

9.1 Country Summary

Nigeria consumes a fairly large amount of natural gas – 116 bcf in 1986, compared to 107 bcf in Thailand.[1] Already by 1980 half of Nigerian industry was fuelled by gas, and gas also contributed over 50 per cent of electric power.[2] However the amount of gas flared each year in Nigeria is higher than in any other country in the world, at 445 bcf.[3] The stated aim of government policy for nearly twenty years has therefore been to reduce flaring, and harness this wasted energy for the Nigerian economy. The irony is that government moves to achieve this aim have had virtually no effect: about 80 per cent of the gas consumed in Nigeria comes from non-associated gas fields, while associated gas is still frequently being flared. Most of the consumers now receiving natural gas are long-standing customers, dating as far back as the early 1960s. Several of the major gas consumers connected in more recent years, as a result of government initiatives, are in fact highly inefficient and their use of the gas is uneconomic.

Three major issues therefore come to the fore:

(a) Can associated gas be substituted for the non-associated gas which is currently being used?
(b) Is it possible to reduce flaring of associated gas, by increasing the size of the market?
(c) Why has the government had such limited success in promoting greater use of gas?

Clearly the large quantities of gas that are a by-product of oil production could not all be absorbed by the Nigerian economy, and reinjection is not economic in most Nigerian oilfields. Some flaring is therefore inevitable. Furthermore gathering associated gas is very

---

[1] Cedigaz, *Le Gaz Naturel dans le Monde en 1986*.
[2] World Bank, internal report.
[3] Cedigaz, *op. cit.*

expensive in Nigeria because the fields are not very large, and the terrain is difficult. There is also the perennial problem of using associated gas, that the reliability of supplies is heavily dependent on oil production. It must however make sense to use associated gas as far as this proves possible, and to save the non-associated fields for future generations. The present pattern of gas usage in Nigeria can be attributed, at least in part, to past government policies and actions.

Essentially the government assumed total powers of disposal over associated gas, with the intention of ensuring that it was used and not flared. This meant that the oil companies had no right to develop markets for the associated gas that they were producing along with their oil. In practice however the government failed to actually carry through projects to use the gas. Recent projects have in fact all been implemented by the oil companies, who were anxious to earn a return on their investment, and were able to demonstrate their goodwill towards the government by supplying one of the latter's favoured projects. Since they have been implemented by the companies they have had to rely on non-associated gas, since the companies have no right of disposal over their associated gas.

The reason why government policies have failed to achieve their aims seems to lie in politicians' preoccupation with political rather than economic issues. In particular the national oil company has had little scope for pursuing its aims unhindered by political interference. A major priority for the government is to maintain stability between the country's different provinces. This is of course a necessary condition of any effective economic policy, and was a pre-eminent requirement in the 1970s, following the country's bitter civil war. The need to appear even-handed however still overrides all other factors. In addition, economic planning is a haphazard affair. It has been described as a process of aggregating a number of projects, rather than co-ordinating and prioritizing them. This approach is not only ineffective in itself, but also provides enormous scope for interference in decision-making by external factors, which would be far more easily prevented in a more coherent framework.

Further utilization of Nigeria's natural gas will require careful allocation of resources according to economic criteria. Only this can create conditions where the country is able to give reasonable compensation to the oil companies for their investment. (At present there are rarely written contracts to govern supply of gas, and prices are sometimes still at 1960 nominal levels.) There is tremendous scope for new schemes; but they are only likely to emerge if the oil and gas industry (Nigerian and foreign) is given more autonomy of wider political goals.

9.2 The Energy Balance in Nigeria

Nigeria is a country well endowed with energy resources. In 1986 it had remaining oil reserves of some 16 billion barrels and gas reserves of 47 tcf;[4] coal and lignite reserves of between 270 and 980 million tonnes; and potential for 8,000 MW of hydropower.[5]

Historically Nigeria depended heavily on the use of coal, with more than half of all energy consumption in the 1950s relying on this fuel. Even in 1967 at the time of the civil war, coal provided as much as 30 per cent of Nigeria's energy. The war badly disrupted coal production in the eastern region however, and it has never since recovered. Meanwhile Nigeria's first oil was produced in 1958, and oil took on an increasingly important role in the nation's economy through the 1970s, with production increasing from 0.54 mb/d in 1969 to 2.26 mb/d in 1974.[6] Export revenues from oil grew even more dramatically over those five years as the price of oil rose from $2.20 per barrel to $14.70. The enormous growth in the oil industry in Nigeria and increasing prosperity from export revenues were coupled with the maintenance of very low oil prices within the country. Between 1974 and 1981 use of petroleum products increased at an annual rate of 18.4 per cent. Of 11.2 mtoe consumed in 1981, 79 per cent was petroleum products, 15 per cent was natural gas, 5 per cent hydro, and less than 1 per cent coal.[7]

9.3 Gas Reserves, Production and Infrastructure

(a) *Reserves and Production.* Nigeria has huge gas reserves, usually put at 47 tcf. (The World Bank and Shell quote proven reserves at nearer 85 tcf, but other sources agree on the lower figure.) This ranks Nigeria sixteenth in the world in terms of proven reserves, and expectations of large undiscovered reserves are also high. As in most countries there has been no exploration specifically for gas, and all those fields so far discovered have been found when looking for oil.

Consequently the known gas fields all lie in the major oil-producing area of the Niger Delta. This is about 225 miles from Lagos, the main potential market. While not a prohibitive distance, it is none the less significant. The fields are widely scattered over the highly faulted swamps of the Delta area, and are mostly quite small, leading to relatively high development and gas-gathering cost.

---

[4] *BP Statistical Review of World Energy*, 1987.

[5] World Bank, internal report.

[6] NNPC, *Annual Statistical Bulletin*, 1983.

[7] World Bank, internal report.

One of the most significant features of Nigerian gas is that about half of known reserves are associated with oil. This has given a high political profile to the issue of natural gas utilization, because in the absence of a gathering system large amounts of gas are flared – a highly visible apparent waste of a resource. During the 1970s as much as 95–98 per cent of all gas production in Nigeria was flared. By 1985 this had fallen to 74 per cent: but only a quarter of the reduction was due to reinjection. Most of the savings were due to the reduction in oil production (1.3 mb/d compared to 2.3 mb/d in 1979). Flaring still accounted for 1,258 mcf/d of gas, more than any other country in the world.

In 1986 total gas production was 643 bcf. Of this total 445 bcf were flared, 81 bcf reinjected, and just 116 bcf actually marketed.[8] 80 per cent of the gas actually consumed came from non-associated fields.[9] Despite the apparent waste of flaring associated gas it should be remembered that the reserves/production ratio is still very healthy, at over 100 years.[10] Some flaring is probably inevitable; but this needs to be kept under constant review.

(*b*) *Infrastructure*. The gas pipeline system in Nigeria is extremely limited. Through the 1960s and 1970s Shell built several short lines to supply local customers in the Delta region. In 1982 the Nigerian National Petroleum Corporation (NNPC) commissioned the Aladja associated gas gathering system to supply the Delta steel plant, and in 1984 the pipeline from Oben to the Ajaokuta steel mill. But in the Aladja system NNPC experienced severe operational and maintenance problems and the Ajaokuta steel mill is still to be completed at the time of writing (January 1987). NNPC planned to complete some smaller pipelines in the Eastern Delta by the beginning of 1987: one would bring non-associated gas from the Alakiri field to the Onne fertilizer plant, another would augment gas supply to the Afam power station from the non-associated gas field Obigbo North. In addition repairs and improvements to other parts of the eastern gas system were planned. West of the Niger, NNPC is about to construct a 220-mile 36-inch pipeline to transport gas from the Utorogu non-associated gas field via Warri to the Egbin power station near Lagos.

The major gas-producing area in the Delta, and the major potential market in Lagos, are still not linked by a pipeline. A contract was awarded to Saipem and Snamprogetti in 1983 to build a 240-mile line

---

[8] Cedigaz, *op. cit.*
[9] World Bank, direct communication.
[10] *BP, op. cit.*

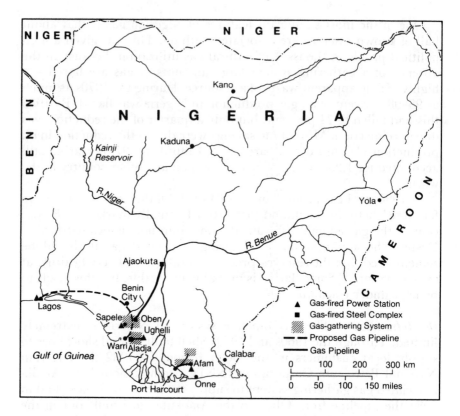

from Warri to Lagos, but work is only due to start on it in June 1987, with completion expected for October 1988. The reasons for delay have been twofold: the downturn in Nigerian revenues caused by falling oil prices, and a freeze on loans for the $250 million project. While these constitute valid reasons for a delay it should be remembered that in the mean time the Egbin power station, designed to use gas from the pipeline, has had to go on stream burning fuel oil.

Recently it seems as though progress may have been made on the project. Costs have been cut, by reducing the scale of the scheme. Instead of an extensive network of pipelines gathering gas from associated fields in Bendel and Rivers states, there may now only be 22 miles of pipeline from a non-associated gas plant at Utorogo (being built by Shell) to Warri. Financing worries also look open to solution, as described in Section 9.8.

9.4 The Government Sector

A constant problem in Nigerian energy policy-making has been the predominance of political over economic considerations. In particular the need to maintain political stability between the regions of Nigeria has frequently distorted economic judgements. Superimposed on this basic problem has been a tendency for the government to make decisions according to short-term political expediency, and a balancing of interest groups, rather than following long-term plans for best use of the nation's resources. The independence and autonomy of the professionals in the national oil company has been drastically undermined, resulting in low morale and weaker performance than might be expected.

Two examples of regional pressures can be quoted. First, in the early 1960s an effective lobby in Imo state prevented development of a non-associated gas field at Akukwa near Enugu. Preference was given to coal-mining in the state despite the higher costs involved. Right through to the 1970s coal was given higher priority than gas in this area, and to this day NNPC has formulated no serious gas project to take gas to the industrial cities of Nigeria's eastern provinces. The second example is the recurring idea that a gas distribution network should supply all the regions of Nigeria, and at equal prices. Given the large distances involved, the variations in economic development and hence market potential, and the enormous capital investment required, this policy is not very realistic. Its only justification is to avert any criticism of favouring one region of the country over others. Indeed looking at the country as a whole one finds that almost all major investment projects are well distributed over the northern, southern and eastern regions of the country. This applies to the two steel plants and three rolling mills, the three refineries, the two petrochemical plants and the fertilizer plant, the petroleum product depots (one for each state) and to a slightly lesser extent to power plants.

In balancing regional interests physical location of industrial plants appears to be more important and more effective than ownership participation of regions or states. For example in 1962 the federal government was given the right to obtain up to 20 per cent and the three regional governments of the western, eastern and northern region each 10 per cent of the shares of the Port Harcourt refinery. Such financial participation by the regions did not turn out to be a stable solution. The refinery is now owned by NNPC and two new ones were built, one in the north and one in the south of the country to 'balance' the existing eastern refinery.

While it is easy to criticize the apparent illogicalities of such planning practices, however, it must be remembered that successive governments have managed to balance the regional demands in the country, and to maintain some political stability. This is no small achievement following the disastrous civil war of 1967.

(a) *Nigerian National Petroleum Corporation (NNPC)*. In 1971 the government formed NNOC, the Nigerian National Oil Corporation, as a vehicle to acquire participations in the various petroleum concessions. In 1977 NNOC become NNPC, the Nigerian National Petroleum Corporation, after merging with the previous Ministry of Petroleum Resources. Exercising the government's authority in the oil and gas sector NNPC thus become responsible for the collection, treatment, transmission and distribution of natural gas. It could take associated gas for free at the flare and non-associated gas at the field from the relevant joint venture for a price based on the crude oil tax regime.

NNPC's capability to deal with gas issues has been slowly developing since about 1978. First some thirty technicians were trained to manage the new facilities built by Shell at Sapele. In 1980 a specialized gas unit was set up in the commercial division of NNPC. In the course of a general reorganization of NNPC in 1986 a full gas division was set up and its headquarters moved from Lagos to Benin City, together with the exploration and production division of NNPC. There is also a separate unit dealing with the proposed LNG scheme. These changes in the institutional framework are intended to strengthen NNPC's capacity to manage existing gas use and to plan new projects.

In the past however NNPC's performance has been severely limited by its powerlessness to take policy initiatives. Its activities have been circumscribed by government instructions. For example it was required to build gas supply systems for the power station at Aladja and the steel plant at Ajaokuta. Close analysis of the prospective customers would have revealed that their gas demand requirements were overstated, and they were unlikely to be able to pay for the gas. NNPC gas staff had no incentive to seriously investigate the quality of the investments, however, since these were priority projects for the government, which NNPC was simply instructed to implement. Similar pressures are now being applied on NNPC to supply gas to a fertilizer plant at Onne, without full feasibility studies being carried out.

It is notable that between 1970 and 1980, when the government was ostensibly encouraging greater use of gas, no money was in fact allocated to such development. Under the second and third development plans NNOC/NNPC invested primarily in acquisition of joint-

venture participation, own exploration, refining and petroleum product distribution. The only investment in gas completed in the 1970s,
the small Sapele supply system, was both constructed and financed by
Shell at NNPC's request.

It is not surprising that NNPC staff have grown demoralized and
undynamic under these circumstances. As a result the company has
experienced severe operational and maintenance problems in all its
activities – refineries, product pipeline and depot system, and associated gas-gathering facilities. An NNPC staff paper attributed such
shortcomings in the case of the Kaduna refinery to 'poor human and
material management, shortfalls in production, infrequent electricity
supplies and poor maintenance of the depot facilities, low price level
of some white products and non-standardization of facilities (e.g.
cylinders for LPG), lack of managerial foresight, adulteration, industrial actions etc.'[11]

Serious deficiencies in NNPC's management were identified by the
Irekefe Tribunal, which was set up to investigate allegations about
large-scale corruption in NNPC in 1978. And NNPC's former Managing Director, L. Amu, acknowledged that 'the poor manpower development in the oil industry cannot be wholly blamed on foreign
companies. Nigerians have their share of the blame, in their attitude
to work and lack of national commitment.'[12]

Despite all the criticism that has been levelled at it, however,
NNPC is neither inactive nor incapable. The company has implemented large refinery projects at Warri and Kaduna, as well as a
1,875-mile petroleum products pipeline system and several petrochemical projects. In response to the Irekefe Tribunal NNPC also
improved the quality of work in its Finance and Accounting Department: between 1982 and 1985 it brought an almost non-existent
accounting system up to date. Given sufficient autonomy and an
allocation of funds there is no inherent reason why it should not install
a comprehensive and well-planned gas distribution system. Its major
problem lies in acquiring the commercial autonomy to pursue such a
policy.

9.5 The Private Sector

Up to 1958 only 'companies which had their principal places of
business in Britain or its dominions and whose Chairman or whose
majority shareholders and directors were British subjects' were

[11] NNPC, *Problems in Petroleum Products Distribution in Nigeria*, 1984, p. 24.
[12] Amu, L. A., *A Review of Nigeria's Oil Industry*, Lagos, 1982.

allowed to explore for oil in Nigeria. This provided Shell/D'Arcy (Shell/BP after 1956) with a virtual monopoly on exploration in Nigeria. However, when oil production started in 1958 the first non-British company, Mobil, started exploring and was followed in 1960 by Amoseas (now Texaco/Chevron), Gulf Oil and Tennessee (now Tenneco). In 1962 AGIP and SAFREP (noe Elf) joined them, followed by Phillips in 1964. By 1964 all the foreign companies producing oil today as joint-venture partners of NNPC were active in Nigeria, with the exception of Pan Ocean which entered Nigeria in 1970 through a farm-out from Delta Oil, the first and only local private company to enter the oil business (in 1966).

(a) *Problems over Associated Gas.* Under the original concessions gas discovered by companies was theirs to flare or market as they chose. They were free to negotiate gas sales contracts with private and public customers, and were also allowed to own and operate gas infrastructure. However, as described earlier up to 95 per cent of associated gas was flared under these arrangements. The government was understandably concerned at this, and brought in several measures to try and limit unnecessary wastage of gas. First a confusing amendment to the Petroleum Act was passed in 1973, which appeared to remove the companies' ownership of associated gas; it certainly required government approval of any future gas price agreements. In 1975, this was followed by the attribution to NNPC of all rights over gathering, transmission and sales of associated gas. The government was henceforth to take such associated gas free of charge at the flare. This removed the possibility of companies making any return on associated gas, and left them with responsibility only for non-associated gas.

NNPC did not have the financial or technical capacity to utilize the quantities of associated gas that now became its responsibility. Schemes to export the gas as LNG, or to build methanol or fertilizer plants, were suggested by the oil companies. Under their new role this was to prove one of their major contributions: presenting utilization plans, with detailed cost estimates and economic analyses. To the frustration of the companies none of the suggested projects came to fruition however, as discussed in Section 9.7.

In 1979 the government introduced Decree 99 in a last effort to eliminate flaring by 1984. Shell, Mobil, Ashland and especially AGIP all introduced reinjection programmes, partly in response to this Decree. But many oil reservoirs in Nigeria do not need pressure enhancement through gas reinjection to increase or maintain oil production (e.g. due to good water drives). Reinjection of gas for conservation purposes alone would often be prohibitively costly. In 1985 the government therefore supplanted Decree 99 by a system of penalties for flaring. The penalty is very low, at 2 Kobo (about $0.005) per mcf of flared gas. Some reinjection still occurs, where reservoir characteristics make it appropriate (for example AGIP's projects at Obiafu-Obrikam and Kwale-Okpai). But, as stated in Section 9.3, in 1986 only 81 bcf were reinjected, compared to 445 bcf flared.

(*b*) *Gas Sales by the Oil Companies*. Following the first discoveries of gas the companies succeeded quite quickly in establishing a market. The first gas sales contract was concluded in 1960 between Shell/BP and the Electricity Corporation of Nigeria. Gas started flowing in 1963 and by 1966 most of today's consumers were connected. On the eastern side of the Niger Delta several small industrial users, the Port Harcourt refinery and the electricity plant at Afam were connected. In the western Delta, near Ughelli, the Delta power station and a glass factory were early customers. Shortly after the civil war, in 1970, Shell/BP also started delivering gas to an industrial customer at Aba (Inter Equitables) thus honouring a contract concluded before the war in 1966. These various facilities were supplied by a mix of associated and non-associated gas sources.

The government policy in the 1970s of removing companies' rights over associated gas unfortunately led to stagnation in market development. Shell no longer actively sought new customers and was also reluctant to accede to NNPC's requests for new investments. On the one hand they were worried by the unclear contractual situation, and were probably disturbed by the poor economic justification for some of NNPC's favoured projects. On the other hand however, they must also have been keen to retain their position as the only oil company supplying gas to the domestic market, in view of the potential for future developments in the gas market.

Some scattered developments have taken place in recent years, driven by the government investment programme in power, steel and fertilizers. For example in 1978 Shell built facilities for NNPC at the Sapele non-associated gas field to supply the local power plant. At the same time another glass factory near Ughelli gained access to gas. Oben, Obigbo North and Utorogu fields were also developed to serve steel and fertilizer plants. Two points should be noted about these schemes. First, they nearly all use non-associated gas (which is still indisputably the oil company's property) rather than the associated gas which is often being flared under NNPC's jurisdiction within close proximity to the non-associated gas field. This is clearly far from ideal. Secondly, most of these activities are undertaken in a legal and contractual vacuum, based merely on exchanges of letters and verbal understandings reached with NNPC, or directly with the customer. For example Shell supplies Delta Glass with up to 2.2 mcf/d, but no formal agreement on quantities, prices or billing arrangements has ever been signed. Many other formal agreements expired in the 1970s, but gas continues to be supplied under the original contract terms.

Due to a series of inefficiencies and uneconomic investments neither the steel plants, the fertilizer company nor NEPA (the power authority) is able to pay a gas price high enough to attract the oil companies' interest. Nor has the government taken any visible steps to remedy the situation. Consequently the companies maintain only a low level of interest in promoting gas sales for the internal market, although they still hope that an LNG scheme may prove possible. Meanwhile they invest in occasional reinjection programmes if these look financially attractive, in order to maintain goodwill with the government.

9.6 Gas Policy, Contracts and Pricing

During the 1967–70 civil war the military government took steps to achieve greater control over both the oil companies as such and over petroleum product distribution and pricing (Petroleum Control Decree of 1967 and Petroleum Decree of 1969). Part of the government's motivation at that time was to control the allegiance of the oil companies – for example SAFREP originally supported the secession of Biafra. Following the rise of OPEC, which Nigeria joined in 1971, the government transformed the structure of the industry. Through the 1969 Petroleum Decree it extended its power to control markets and it started acquiring participations in the oil companies' operations through the newly formed Nigerian National Oil Corporation (founded on 1 April 1971). By 1974 NNOC had acquired a 55 per cent share in most companies, which was increased to 60 per cent for all companies in 1979. The 1979 nationalization of BP's share (due to alleged oil trade with South Africa) brought NNPC an 80 per cent share of Shell's operations, which account for about half of all oil production in Nigeria.

The Petroleum Decree defined petroleum to comprise natural gas, thus making gas subject to the same tax regime as crude oil. An amendment to the Decree in 1973 empowered the government 'to take natural gas produced with crude oil by the licensee or the lessee free of cost at the flare or at an agreed cost and without payment of royalty'. It also obliged 'the licensee or lessee to obtain the approval of the Federal Government as to the price at which natural gas produced by the licensee or lessee (and not taken by the Federal Government) is sold'. This wording was extremely ambiguous as to the ownership of associated gas, and has been the cause of disputes ever since its publication.

The next major step in government gas policy came in 1979, with Decree 99. This ordered that all flaring of gas was to stop by 1984, with gas being either reinjected or sold. This decree was based more on wishful thinking than on realistic plans. Reinjection was not a viable solution for many of the wells; and the prospect of absorbing such large amounts of gas into the economy in a period of just five years were slim. When 1984 arrived and extensive flaring was still continuing the government amended the decree, to allow flaring where the field was remote, the gas of low quality, or where similar problems were encountered. The need to be 'reasonable' was admitted by the Minister. He also pointed out that the decree had been introduced at a time when oil was selling per $40 per barrel, and was expected to rise to $50. The realities of 1984 were very different to the expectations of both government and companies.

A problem in recent years has been the lack of a clear energy policy, with most decisions being handed down direct from the military, based on political rather than economic criteria.

(*a*) *Contracts*. During the 1970s the government made several moves to increase its control over the oil companies. Since 1970 concessions have no longer been granted. The only company to have been awarded a production-sharing contract since that time has been Ashland Oil, in 1973. By 1979 risk service contracts had become the favoured vehicle for oil exploration and production. Elf, AGIP and Nigus Petroleum were all awarded such contracts in 1979, but have not yet started production under them. Since that time in fact oil production has been steadily falling under OPEC agreements (from 2.1 mb/d in 1979 to 1.3 mb/d in 1985). Under these circumstances no further contracts have been awarded.

(*b*) *Pricing*. Producer gas prices in Nigeria can fairly be described as anarchic. Gas prices negotiated in the 1960s under contracts with Shell/BP have been left untouched and continue to be respected by all parties. Prices remain at the level of the 1960s in nominal terms (0.13–0.22 Naira per mcf or $0.03–0.06/mcf at today's exchange rate). Escalation clauses contained in gas price contracts between Shell and its customers have never been invoked. However gas supplies to local customers are about to be taken over by NNPC, and new arrangements will then be introduced.

More recent gas delivery arrangements have been based on informal understandings between NNPC and Shell. Shell receives about

0.45 Naira per mcf for gas delivered at Oben and Ughelli (for Delta Steel) (around $0.70/mcf in 1981, $0.12/mcf in 1986). Negotiations over firm contracts have been lengthy and inconclusive, for example discussions between NNPC and Shell about gas sales from the Oben field have been going on since 1977.

NNPC in turn has not been able to agree on prices or contract terms with its customers, NEPA and the steel factories. Essentially these companies do not want to pay, and probably are not in a position to pay, all their suppliers. NNPC therefore simply bills them at prices which have never been formally agreed. For it is not an option for NNPC to stop delivery to government-owned customers.

In 1984 NNPC developed proposals for a national gas pricing policy, but these are still under discussion today. The issue of differential pricing between the provinces (based on their distance from the gas fields) is proving particularly difficult to resolve. An inter-ministerial committee, with representatives from various ministries and from NNPC is currently working on a new consumer pricing policy.

The comparative pricing of various fuels is also of course an important component of any new gas pricing policy, and in itself is a thorny problem. So far gas has remained competitive with other fuels in most markets, thanks to the retention of 1960s nominal prices, despite the very low prices which prevail for fuel oil. Since 1975 the government has controlled domestic petroleum product prices in Nigeria. From 1975 to 1978 petroleum prices were set close to their border price equivalent, but with the exception of fuel oil which was priced at only about 40–65 per cent of its border price equivalent. The government adjusted petroleum product prices in 1979 following the increase in international prices. But while gasoline almost reached international levels, particularly after a second readjustment in 1982, fuel oil, the main substitute for natural gas, lagged behind international prices despite an increase in the price of low-pour fuel oil by over 100 per cent in 1979. High-pour fuel oil remained at around 20 per cent of the international price and low-pour fuel oil at between 40 and 50 per cent.[13]

While there is no standard gas price, World Bank estimates show that in general gas has been competitively priced compared to high-pour fuel oil in the Niger Delta area, close to the gas fields. Even in Lagos it would have been competitive with low-pour fuel oil (but not high-pour fuel oil). Industrial and commercial users located close to the trunk line would therefore find it in their interest to use gas rather than low-pour fuel or diesel. This is borne out by the actions of

---

[13] World Bank, internal document.

potential consumers. A cement plant close to the Ajaokuta–Oben pipeline requested access to the gas in 1984 but it is not known whether NNPC followed this up. The Delta Glass company actually financed a spur line to the Ughelli system and agreed a price with Shell in 1981, based on an exchange of letters.

The irony is that relatively so few customers have been linked to the gas supply system, given the fact that current pricing would make this advantageous from the consumer's point of view. It is said that several potential customers have enquired with NNPC about connection to a nearby pipeline, but have never received a reply from NNPC. It seems that there is unsatisfied demand, and market opportunities are being lost. In the longer term however it is clear that a policy needs to be formulated which will allow formal sales contracts to be agreed, and prices will have to be increased if the producing companies are to be induced to sell their gas. This will need to be part of a comprehensive restructuring of energy prices in order to maintain the competitiveness of gas against other fuels. While formidable internal political problems surround this issue there seems little hope of further gas utilization until it is resolved.

9.7 Markets

The overall picture of gas use in 1984 is given in Table 9.1. These figures tell their own story, both in terms of the proportion of gas that is flared, and in the enormous relative importance of the power sector as a gas consumer. Arguably this sector could absorb a much larger amount of gas: but historical preference for hydro schemes preempted the construction of thermal power stations.

One of the major results of early failure to capture the power

Table 9.1: Nigerian Gas Consumption. 1984. Million Cubic Feet per Day.

	Volume	%
Flared	1,239	79
Reinjected	56	3
Oilfield Use	29	2
Refineries	2	–
Electricity	219	14
Industry	26	2
	1,571	100

Source: World Bank, direct communication.

market has been a twenty-year delay in building a pipeline to the major industrial markets for gas in the Lagos area. This project now at last seems close to becoming reality, and will open the way for a potential doubling of the industrial market for gas, as well as increasing the uptake by the electricity authority. A small potential for gas also exists in water heating and air-conditioning of hotels, offices and richer homes in Lagos. This sector would be unlikely to use a significant amount of gas however. Meanwhile some independent studies are being undertaken of the potential for compressing small amounts of associated gas, to be used in villages close to gas fields to run generators and small industries.

To completely eliminate gas flaring will remain impossible however, assuming continued oil production, since a large enough market will not exist in the foreseeable future. An imaginary alternative scenario, in which gas is used in all available (and economically beneficial) markets shows this very clearly. If the major hydroelectric scheme at Kainji had not been built, but instead a 760 MW gas power station had been built at Lagos, this would consume about 90 mcf/d (running at 50 per cent capacity). A pipeline would then have been built to Lagos, which could also have supplied the cement factories at Shagamu and Ewekoro, and firms in the Ikeja industrial park in Lagos. If existing uneconomic projects like the steel factories were excluded, total net additional consumption would still only be about 150 mcf/d. This would represent an increase of some 62 per cent over 1984 actual consumption; but flaring in that year would still have amounted to 1,100 mcf/d – around 65–70 per cent of total gas production.

Export schemes such as LNG offer a further route for gas utilization, and one that is favoured by the oil companies since they would earn hard currency. Some good opportunities for such schemes have been missed in the past. The LNG option is now being explored once again.

(a) *Power.* As in other developing countries extensive development of gas reserves needed a bulk market, which could most easily be found in the power sector. Early decisions in Nigeria to favour hydroelectric projects over gas-fired power stations effectively delayed construction of a pipeline to Lagos by twenty years. This in itself might have been justified if hydropower was in fact cheaper than gas. Several studies have consistently shown this not to have been the case however, so an alternative explanation for the government's preference for hydro-electricity is needed. In 1953 a hydrological investigation of the Niger river was started, which in 1959 resulted in the recommendation to develop hydropower sites at Kainji (then named Kurwasa), Jebba and Shiroro. After oil and gas were discovered and production started in 1958, studies by the Electricity Corporation of Nigeria easily established that gas-fired power was competitive with coal-, fuel oil- or diesel-fired power stations.

Meanwhile the World Bank, which had been asked to finance the Kainji hydro project, commissioned a study of the relative advantages of a hydro-based power expansion plan and the alternative of smaller gas-fired plants (which Shell/BP was promoting). Proven gas reserves at that time (1963) already amounted to 3 tcf, with probable reserves of 10 tcf. The World Bank's conclusion was that the hydro plan was in fact competitive with gas up to a 6.5 per cent discount rate on capital investment. This opportunity cost for capital now seems severely understated, but at that time was acceptable. The Kainji hydro project therefore went ahead.

The economic advantage of gas over hydropower has nevertheless been reaffirmed in numerous studies over the years. Most recently a 1984 study carried out for NEPA (the power authority) favoured gas over all other fuels for electricity generation. To some extent the government has heeded such advice, and several gas-fired power stations have been built. For example in the 1970s two major thermal stations were built at Sapele and Lagos (Egbin) in addition to expanding existing plants at Ughelli and Afam. Sapele started taking gas in 1978/9. The first 220 MW unit at Egbin was completed in 1985, but NNPC is still trying to get construction of a pipeline to Egbin under way. When the first of six units came on stream in 1985, no gas was available, and it has had to be run instead on exportable fuel oil. A gas supply to the power station now seems unlikely before 1988.

By 1984 the power sector was consuming 219 mcf/d of gas, more than double the amount used in 1979 (95 mcf/d). This represented 14 per cent of all the gas produced in Nigeria in 1984.[14] But at the same time old hydro plans were still being put into effect. The two plants which were originally studied in 1959 along with Kainji, have been built in the last few years. Jebba was finally completed in 1985, and Shiroro is just nearing completion. These decisions cannot be justified in economic terms and must be attributed to a combination of inertia and vested interests, presumably coupled to the willingness of financing agencies to promote them.

(*b*) *Industry*. Industrial customers include three glass factories, textile mills, and steel mills. While total industrial consumption only

[14] World Bank, direct communication.

amounted to 26 mcf/d in 1984, about 2 per cent of gas production, this represented 50 per cent of total industrial energy use.[15] NNPC has not played an active role in encouraging more firms in the Delta area, or along the Ajaokuta pipeline, to connect to gas. On the contrary, when companies have asked to be connected they have had no apparent response from NNPC. Consequently there is still scope for further penetration of gas into the industrial sector, even where pipelines already exist. The construction of pipelines to other areas would of course extend this potential even further, particularly in the case of Lagos, where several industrial estates would be served.

The projects actually undertaken by NNPC however have been fraught with difficulties. Clear examples are the Aladja and Ajaokuta steel mills, which are the most expensive gas projects yet undertaken in Nigeria. The 24-inch line from Oben to Ajaokuta is about 125 miles long and was commissioned in 1984. The Aladja system (commissioned in 1982) gathers associated gas via seven compressor stations and has a total length of about 70 miles. Together these projects probably cost over $200 million. Yet the Ajaokuta steel mill is not yet completed and may never be able to cover its operating costs. The Aladja system has barely functioned, due to operating and maintenance problems and a lack of spare parts. Offtake by the Aladja steel plant, which is mainly supplied from the NNPC/Shell field at Ughelli East has never reached target volumes and is fairly erratic. A fertilizer plant is now planned for Onne, which will use 50–60 mcf/d. But it is far from clear that this will be economic.

(c) *LNG*. Domestic gas development has been continuously over-shadowed in public discussion by a series of proposed LNG schemes. The first such scheme was prepared by Conch, Shell/BP and British Gas in the early 1960s and was almost signed by the government in 1965; but BP discovered large quantities of natural gas in the North Sea and the project was abandoned. According to industry sources the project could have been signed and agreed had it not been for a year-long delay in government decision-making in Nigeria caused by inter-regional differences of opinion. However, it remains an open question whether such an agreement would have survived the North Sea gas finds, unless construction had already been far advanced.

Since the civil war various LNG schemes requiring between 0.5 and 2 bcf/d of gas have been proposed, debated and sometimes even negotiated. For example Gulf Oil proposed a $1 billion LNG plant of 500 mcf/d capacity in early 1974. In early 1975 the Minister of Mines

[15] *Ibid.*

and Power approved two partnerships for LNG: one with Shell for an LNG/NGLs project at Bonny (1 bcf/d) and one with AGIP/Phillips for a plant of the same size for LNG, NGLs and ethane extraction at Escravos. The various ideas evolved into a joint venture (Shell, AGIP, Elf, BP, Phillips, NNPC) for an LNG project of 2 bcf/d at Bonny. In 1977 negotiations for a gas sales agreement were initiated and basic agreement on price, indexation, take-or-pay and condensate disposal was reached in 1979. An LNG long-term sales contract could have been concluded at that time, but government approval from the new civilian government was not forthcoming, and the opportunity was lost. The $12 billion project never took off, and Bonny LNG Ltd voluntarily liquidated in 1982. Current market opportunities are restricted; but the long-term hope remains that a market may be found for some of the Nigerian gas as LNG exports.

In 1985, a new joint venture was formed. The members are NNPC, Elf, AGIP and Shell, with the latter responsible for the technical development of the project. The project is designed for an initial delivery of 4.5 million tonnes of LNG a year from a two-train liquefaction plant, to be located at Bonny. The start-up of deliveries is scheduled for 1995 but no commitments have as yet been made by any potential importers. One of the questions is whether there will be sufficient demand for additional supplies to Western Europe before the late 1990s.

9.8 Finance

Gas development in the 1960s was financed by Shell/BP. In the 1970s Shell financed the Sapele facilities, and Delta Glass financed its own tie-in to the gas system. Non-associated gas development and gas reinjection is financed under the usual system of joint-venture cash calls in proportion to each party's share in the joint ventures, or under a production-sharing contract in the case of Ashland (the only company with such a contract in Nigeria). NNPC financed its share in such ventures as well as its investments in associated gas gathering and in transmission out of its current cash flow (as authorized under its government-approved capital budget).

Financing was clearly no problem for NNPC until the early 1980s. Between 1969 and 1974 Nigeria's oil revenues increased twenty-eight times, from about $430 million to $12.1 billion per year. During the 1970s and well into the 1980s project expenditure remained well below budgeted targets. While money was available however the government simply did not allocate it to gas development. During the second five-year development plan (1970–74) nothing was allocated to gas investment. In the third five-year plan (1975–80) only 10 million Naira out of NNPC's total 1.757 billion Naira were allocated to gas development. By 1977 nothing at all had been spent. During the fourth five-year plan a total of 16.4 billion Naira were to be spent in the oil and gas sector of which 1.6 billion Naira (9.6 per cent) was for domestic gas development and 3.0 billion Naira for LNG (18.3 per cent). Of this only relatively small amounts were actually spent.

It could perhaps be argued that not enough money was left for gas development after the government's main priorities had been financed. But even during the oil boom, when borrowing was relatively easy for Nigeria, no attempt was made to borrow money for gas development. The first such move was in 1978 when NNPC began discussions with the World Bank over the financing of a pipeline from Warri to Lagos. However, these discussions were not seriously pursued until 1982/3.

The Warri–Lagos pipeline is thus the first gas project where NNPC has tried to obtain third-party financing from export credit agencies and multilateral banks (the World Bank and the European Investment Bank). As yet they have been unsuccessful in concluding a deal however. The World Bank is willing to lend $200 million towards a gas-gathering system and compressor station at Warri, but only on condition that the government radically changes its pricing policy for petroleum products. It also wants a foreign operating company to be involved, since it feels this will improve efficiency. NNPC is reluctant to forgo its independence in this way; but the government has now reduced the subsidy on fuel oil by as much as 30 per cent, which has made the outlook for World Bank agreement much better.

Should World Bank agreement be forthcoming there is a chance that this will then act as a catalyst to free the Italian export credits which are currently frozen. These are intended to cover Saipem and Snamprogetti's construction of the 240-mile pipeline to Lagos. The original offer of funds from the Italian export credit guarantee agency was withdrawn in the wake of Nigeria's failure to come to an agreement with the IMF. The contractors are still keen to carry out the work, and in case export credits remain unavailable they are considering a countertrade deal to overcome the problem.

In the final analysis obtaining finance is not a limiting factor to gas utilization in Nigeria. Obtaining external funding may involve unpalatable policy concessions; but if necessary, and if NNPC decided to give priority to gas development, sufficient funds could be found from national sources. The ultimate blockage on gas use lies in the government's will to pursue such a policy, rather than in problems with the projects themselves.

Sources

In addition to the publications cited in the footnotes, much valuable information for this chapter was obtained from executives in the following companies and organizations: AGIP, Ruhrgas, Shell and the World Bank.

APPENDIX TO CHAPTER 9

Background Data

Population[a]: 99.7 million
GNP per capita[a]: $800
Average annual growth in GNP per capita, 1965–85[a]: 2.2 per cent
GDP[a]: $75,300 million

External public debt[a]: $13,432 million
External public debt as percentage of GNP[a]: 17.8

Total energy consumption[b]: 40.980 mtoe
   Commercial: 26 per cent   Non-commercial: 74 per cent
Commercial energy consumption per capita[a]: 165 kgoe
Commercial energy imports[b]: 1.232 mtoe
Commercial energy exports[b]: 62.721 mtoe

Proven Reserves and Production[c]

	Gas	Oil	Coal	Hydro
Proven Reserves	47 tcf	16 bbls	Unknown[c]	8,000 MW
Production	643 bcf	1.455 mb/d	114 kt	8,800 GWh

Consumption by Sector: Thousand Tons of Oil Equivalent

	Gas	Oil Products	Coal	Total	% Share
Power	1,047	267	12	1,326	14
Industry	1,093	823	73	1,989	21
Transport	–	5,413	2	5,415	56
Households	–	897	6	903	9
Total	2,140	7,400	93	9,633	
% Share	22	77	1		100

Power Sector[b]

Installed capacity: 2,770 MW
Electricity generation (%): Hydro=33, Oil=12, Coal=1, Gas=55

Sources

(a) World Bank, *World Development Report*, 1987, pp. 202–237. Figures refer to 1985.
(b) OECD, *Energy Balances of Developing Countries 1971/1982*, 1984. Figures refer to 1982.
(c) Gas reserves, and oil reserves and production, *BP Statistical Review of World Energy*, 1987. Gas production, Cedigaz, *Le Gaz Naturel dans le Monde en 1986*. Coal and hydro reserves and production, World Bank, internal papers. Coal reserves are put at somewhere between 270 and 980 mt. All figures refer to 1986, except coal production (1981) and hydro production (1982).
(d) World Bank, internal paper. Figures refer to 1980.

10 PAKISTAN

10.1 Country Summary

Pakistan is an important example of how a low-income developing country can benefit from using its gas reserves; and also of key problems which can arise between a government and oil companies if the benefits of such development are not shared in a mutually satisfactory way. Recent attempts to revive the interest of the private sector in the Pakistani gas industry also provide examples which might be followed by other countries.

Gas has been used in Pakistan since 1955, and there is now an extensive distribution network throughout the country. This early progress can be attributed to a fortuitous set of circumstances. Burmah Oil, which already had interests in India, expressed a desire to explore in Pakistan. Their expectations in terms of ownership rights and rate of return were moderate, and allowed for easy agreement with the government. This was partly a function of the period, when expected rates of return were much lower than in later decades. But it was also a reflection of the company philosophy, with Burmah actively seeking government participation of 50 per cent, in a period when Shell, for example, would still expect a 100 per cent concession. Local Burmah executives also had long experience in the country; they knew the government officials and were familiar with government methods. There were thus close and good relations between the company and the government, allowing gas production to get off the ground. Over time, however, both political and economic conditions changed, while the concession terms remained much the same. This led to a slowing of activity, and to problems for energy supply in Pakistan.

For thirty years the government has encouraged substitution of imported oil with indigenous gas, and has achieved a large measure of success. Gas use has penetrated into every sector of the economy, including the residential market. But this has happened due to the maintenance of extremely low prices for gas. While the markets were successfully captured through this strategy, long-term problems were also being caused. The oil companies lost their initial interest in operating in Pakistan. The first large gas field at Sui was developed under a low-returns regime under a cost-plus contract. While this provided a steady income it did not offer any prospect of large profits.

Other companies were not prepared to take the risk of exploring for further fields if there was absolutely no prospect of earning high returns. Even developing known fields was not particularly attractive, since prices were regulated by the government, and were kept very low. At the same time that new exploration and development were becoming rare events, the market was expanding very rapidly due to the low retail price of gas. The predictable consequence was that demand outstripped supply, and gas shortages have become commonplace.

Shortages of gas coupled with load shedding in the electricity sector in the early 1980s led the government to reappraise its fundamental energy policies. These had been predicated on the dual needs of protecting the economy from the effects of high international oil prices after 1973, and on encouraging development of the indigenous natural gas. Furthermore, the government was not blind to the irony of seventeen known gas fields lying dormant, while customers were unable to obtain gas. Some of these fields may in fact be uneconomic, thanks to the quality of the gas or the small size of the reserves. But nevertheless there remains considerable potential for further development.

Recent policy has therefore been to follow a two-pronged attack on the problem. Energy prices are being progressively reformed, to reflect more nearly the true economic value of the various fuels. This process has been helped by the recent fall in oil prices; by 1986 gas consumer prices matched the border price for fuel oil. This process of price reform must also involve a clear assessment of the value of each fuel to the different market sectors: for instance, heavy use of gas in the cement industry has been curtailed. At the same time new contract provisions have been introduced for the oil companies, and this does seem to have had some effect, although the companies are not yet satisfied with the terms on offer. For all new contracts well-head prices for gas are related to the international fuel oil price. There are two problems with the new system. First, the basic principle is that gas should be priced at 66 per cent of the fuel oil level. In the light of recent price falls this does not amount to very much money. Secondly, a discount is negotiated on this 66 per cent for each concession, according to expected exploration costs, distance from market, and so on. The companies are unhappy with this provision. Their major concern is that the government should remain flexible in its approach, and thus able to respond to new situations such as the change in oil price level. However favourable a system may seem when it is first devised, it cannot hope to cover all possible future eventualities. Nevertheless, several fields that have lain unused for many years are finally moving into production, such as Dhodak and Kandhkot. Exploration has also received a great boost from recent oil finds in the Badin area, which have renewed company interest in Pakistan's oil prospects.

10.2 The Energy Balance in Pakistan

Despite extensive use of local gas and oil the energy balance in Pakistan is not very favourable. Local sources now provide two-thirds of all commercial energy consumed, but nevertheless 60 per cent of the country's export receipts are needed to finance energy imports. This is due to the rapid development of the country, which has especially concentrated on quite heavy, energy-intensive industries. Furthermore the historical pricing policy, under which energy was heavily subsidized, resulted in very little effort at conservation. In fiscal year 1985/6 imports of crude oil and oil products amounted to 6.2 million tonnes, and cost $1.6 billion.[1] This heavy drain on the exchequer was exacerbated by the decline in domestic oil output of the early 1980s, caused by depletion of existing fields and a lack of further exploration. Meanwhile, the shortage of gas in the power sector and in industry led to a large increase in demand for fuel oil. The share of fuel oil in petroleum products consumption rose from 14 per cent in 1980 to 29 per cent in 1985. Oil reserves are now put at 139 million barrels: production in 1986 was 50,000 b/d, but demand was much higher than this at 120,000 b/d.[2] Some coal is produced within the country (2.2 million tons in 1986)[3] but 97 per cent of this is consumed in brick kilns. It therefore provides little direct competition to gas in the latter's major market sectors.

Energy consumption in 1985/6 followed the pattern given in Table 10.1. In 1986 gas already provided two-fifths of all commercial energy consumed, but known reserves offer the opportunity of even greater consumption. The government is therefore following a concerted campaign to encourage exploitation of further gas (and oil) fields, in order to minimize the need to import so much energy. Efforts are also being made to encourage greater conservation of energy; and as described in Section 10.6 pricing policy has radically changed.

---

[1] GDC, direct communication.
[2] *World Gas Report*, 3 November 1986.
[3] GDC, direct communication.

Table 10.1: Energy Consumption in Pakistan. 1985/6.
Million Tons of Oil Equivalent.

	Total Energy		Indigenous Energy	
	Volume	*%*	*Volume*	*%*
Oil	6.98	41.7	0.94	8.8
Gas[a]	6.07	36.3	6.07	56.8
Coal	0.98	5.9	0.98	9.2
Hydro	2.60	15.6	2.60	24.3
Others	0.09	0.5	0.09	0.9
	16.72	100.0	10.68	100.0

Note: (a) excluding consumption as feedstock.

Source: GDC, direct communication.

10.3 Gas Reserves, Production and Infrastructure

(a) *Reserves and Production.* Total remaining gas reserves in Pakistan are variously put at between 15.3 tcf (Shell and BP) and 16.1 tcf (government of Pakistan). They are well scattered around the country, but lie in four main areas:

(a) the centre of Pakistan: Sui, Kandhkot, Mari, Pirkoh;
(b) the north: associated gas in the traditional oil-producing region of Potwar: Dhurnal, Toot, Meyal;
(c) an area west of the River Indus in the Punjab: Rodho and Dhodak;
(d) the south: the recently prolific Badin area.

There has been a large amount of exploration in 1986/7, with Union Texas finding several new reserves to the east of Karachi; the first offshore discovery was also made in March of 1986 by the Oil and Gas Development Corporation (OGDC) in the Indus delta.[4] Currently however the largest reserves are those at Sui and Mari, with 7.3 and 3.7 tcf respectively. Both of these were very large fields by any standards, with Sui originally having nearly 10 tcf of gas, and Mari 4.5 tcf.[5] Many other fields are rather small however, making them of marginal worth in economic terms. A further problem with much of the gas in Pakistan is its variable quality.

These intrinsic qualities of the gas fields, combined with the pricing

---

[4] *World Gas Report*, 10 March 1986.
[5] World Bank, internal report.

problems described in Section 10.6, have meant that only a few fields have actually been developed. The first to be brought into production was Sui as early as 1955, and this still provided 82 per cent of total production in 1985. Other gas fields producing in that year were Mari (180 mcf/d), Sari and Pirkoh (70 mcf/d). In addition associated gas was obtained from the oilfields at Meyal, Toot and Dhurnal, to bring total production in 1985 to 1.13 bcf/d.[6]

During 1986 a further field came into production, at Kandhkot, thirty miles south of Sui. Production started with 30 mcf/d, and is already being expanded to 45 mcf/d.[7] Bids for operating the Dhodak field have also been invited. Dhodak has 700 bcf of gas with up to 35

[6] *World Gas Report*, 3 November 1986.
[7] *Petroleum Intelligence Weekly*, 16 February 1987.

million barrels of condensate,[8] but lies in a mountainous and little-populated area, 125 miles north-west of the city of Multan. The Badin field is also due to start producing gas by mid-1988, once a 100 mcf/d pipeline to Hyderabad is constructed.[9] A second phase of development at Pirkoh will raise output of that field to 180 mcf/d.[10] There is thus considerable activity, with numerous new facilities increasing the availability of gas.

Meanwhile it has been reported that China is considering building a 1,250-mile gas trunk line to Pakistan. This would provide an outlet for the large volumes of gas being found in the Xingjiang Uygur Autonomous Republic and Tingshai province.[11] However it seems extremely unlikely that this will materialize in the foreseeable future.

(*b*) *Infrastructure*. Given the size of the market that has been developed in Pakistan (see Section 10.7), it is not surprising to find an extensive gas pipeline network. As of March 1985 there were 2,437 miles of high-pressure transmission lines and over 8,125 miles of distribution lines in seventy-eight major towns and cities.[12] The two original pipelines to the north and south of the country (to the cities of Multan and Karachi) were planned by the US consultant Cornelius Stribling Snodgrass, on behalf of Burmah. Burmah put in 25 per cent of the equity for the Karachi line, and also built it for the government on schedule – which did no harm to their image in the country. Burmah had a (smaller) equity interest in the much larger northern pipeline, for which they were managing agent. In the 1960s they sold their shareholdings in the pipelines, apart from a residual interest in the northern line, and handed over operation to the local companies. This was a reflection of changing politics and increasing local skills.

Despite a good start, and long total mileage, the development of the network as a whole has been somewhat haphazard, and there is a need for overall strategic planning. Furthermore the original lines have now been used heavily for up to thirty years, and are nearing the end of their useful lives. It is very important that the infrastructure be well planned, since this can make a significant difference to the speed with which additional fields can be brought on stream. It has been suggested that the seasonal problems in meeting peak demand for gas in Karachi should be solved by providing gas storage near to the city, rather than through additional pipeline capacity. No geological

---

[8] GDC, direct communication.
[9] *World Gas Report*, 19 June 1987.
[10] *Cedigaz*, 11 June 1987.
[11] *World Gas Report*, 30 January 1987.
[12] Ahmad, 1985, p. 6.

studies have yet been carried out however, to determine whether there are suitable rock strata.

As described in Section 10.4, the pipelines are now administered by two major companies serving different sections of the country. In addition there are dedicated lines from Sui and from Kandhkot to the new Guddu combined-cycle power station and Mari fertilizer plants. As new gas fields come into production there is a need for new pipelines to serve them, and a number of programmes are currently under way to fill these gaps. For example an additional 170 mcf/d pipeline is being constructed to run from Pirkoh to Sui. The World Bank has funded five separate projects for expanding the northern network; and the ADB is funding an expansion of the Indus Right Bank pipeline, which will bring new capacity to Karachi. A 68-mile pipeline is also to be built from the Badin field to Hyderabad.

10.4 The Government Sector

(a) *Ministry of Petroleum and Natural Resources*. This is responsible for the development of oil, gas, coal and non-fuel mineral resources. The Ministry is composed of four directorates: Petroleum Concessions, Oil Operations, Gas Operations and Energy Resources. The Petroleum Concessions Directorate is responsible for negotiating and granting concessions to exploration companies, including the national oil company and joint-venture companies, and for monitoring the progress of exploration. The Oil and Gas Operations Directorates manage pricing policy and fuel transportation and distribution. The Directorate of Energy Resources is responsible for the compilation and analysis of historical data relating to energy supply and consumption.

Historically the bureaucracy earned the respect of oil companies for its fair handling of their interests. In particular, the government has always been meticulous to pay all amounts due. Over time however the accumulating number of rules and regulations to be followed, and precedents to be observed, has gradually stultified the system. The very meticulousness which used to be praised has now become a problem to companies anxious to extract quick decisions from the government. Where in the early days of independence key civil servants had considerable decision-taking power, this is now checked and balanced by other factors.

(b) *Oil and Gas Development Corporation (OGDC)*. OGDC was established in 1961 as a fully government-owned company, with a mandate to explore and develop the country's oil and gas resources. It was not easy for it to fulfil this task, since its staff were lacking technical experience and funds were always short. This situation improved after 1975, when it was reorganized and granted a larger budget. However criticism of its efficiency is still widespread, both on the managerial and technical levels. A major reason for its shortcomings is that the Corporation is run very much as a government department, rather than as an autonomous business enterprise. In particular it is subject to civil service rules on salary scales, hiring of staff and procurement of materials. The resulting bureaucracy is very unhelpful. Furthermore, there are several other public companies operating in the oil and gas spheres, but not subject to the same salary restrictions, which has meant a constant drain of qualified and experienced staff to competitors. This is important because it is one of the factors which discourages foreign oil companies from activity in Pakistan; a joint venture with a badly-run and staffed national oil company can seem a very unattractive prospect. However there are now plans to put OGDC on a more commercial footing, without annual government funding.

(*c*) *Sui Gas Transmission Company (SGTC)*. This company was set up with finance from Burmah Oil, the Commonwealth Development Finance Company, the Pakistan Industrial Development Corporation and the government. It is now 87 per cent owned by the government however. It operates two pipelines transporting gas from Sui to Karachi, where it sells gas to a separate government-controlled distribution company, the *Southern Gas Company* (SGC, formerly the Indus Gas Company and the Karachi Gas Company). Unlike OGDC both gas transmission companies are self-governing, and have earned reputations as efficient and well-respected bodies. SGTC and SGC are in the process of merging: the one company will then handle gas purification, compression and transmission in Sind and Baluchistan provinces.

(*d*) *Sui Northern Gas Pipeline Limited (SNGPL)*. Over 90 per cent of equity in this company is held by the government. It operates pipelines from Sui to the northern area of Pakistan, with a total pipeline capacity of 380 mcf/d. Unlike SGTC it is also responsible for distribution of the gas, in some fifty towns and cities. Altogether it has some half a million customers.

10.5 The Private Sector

The major companies currently active in Pakistan are: Pakistan Petroleum Ltd. (PPL), Pakistan Oilfields Ltd. (POL), OGDC, Union

Texas and Occidental Petroleum. While exploration is carried out under joint-venture agreements with OGDC the production of oil and gas is primarily the responsibility of the private sector.

PPL has the longest history in Pakistan. As operator of the Sui field it plays a central role in the gas economy of Pakistan. It was founded in 1950, with a 70 per cent stake from Burmah Oil and 30 per cent from the government. Over the years shareholdings have changed hands several times, with the introduction of 12 per cent equity from the IFC, and 1 per cent from local private investors, but overall Burmah Oil always maintained the majority share. During 1986 Shell began negotiations to buy out Burmah, but their bid was ultimately rejected by the government.

The second-largest gas field, Mari, was originally discovered and developed by Esso. The gas from this field has always been marketed direct to a local fertilizer complex, and in 1983 Esso sold out of the operation. 40 per cent of the equity is now owned by the Fauji Foundation (a military pension fund, hence semi-government), with a further 40 per cent held directly by the government and 20 per cent by OGDC.

POL is 60 per cent owned by Attock Oil, and has been active in Pakistan since 1915. It operates the Meyal oilfield, from which associated gas is produced. OGDC operates the Toot field, which also has associated gas.

Union Texas and Occidental have been carrying out extensive exploration since 1977 in the Badin area, 150 miles south-east of Karachi. They have made a number of finds, including several gas fields and some small oilfields. Phillips Petroleum has recently acquired a 50 per cent interest in an exploration block immediately to the west of Badin.

The history of Burmah's success in launching a gas industry in Pakistan is of great interest, because it is unique. The reason for its success seems to lie in the combination of Burmah's relatively modest demands, and the nature and depth of oil company links with the government. Importantly, many of Burmah's local staff had grown up in the country alongside their government counterparts; they spoke the language fluently, and were thoroughly familiar with government practices and procedures. This generated a very high level of trust between the two sides. Furthermore, Burmah had already established an exploration company in India with 50 per cent government participation, in an era when 100 per cent oil company concessions were normal. It originally offered the Pakistani government a similar half share; but as it happened the government was preoccupied at the time with other matters and chose not to take up the share. Instead it offered the share to local private investors, but take-up was very low indeed. The government finally took a 27.5 per cent interest, but did not play a very active role. Nevertheless the offer of a 50 per cent share must have contributed to goodwill. The net result of this happy relationship was that Burmah executives were trusted and their opinions highly valued by the government. They therefore had an indirect role in forming the country's gas policy – as opposed to a more combative relationship where companies are sometimes fiercely at odds with the national policies.

Burmah's required rate of return seems low from today's perspective – originally just 10 per cent. This was not especially unusual in the context of the 1950s however, in an era before oil price rises and large inflationary pressures. The problem with their pricing agreement was the lack of an effective review clause. Again this was not unusual for the time – early British Gas contracts for the North Sea suffered a similar defect. The difficulty was that renegotiating the clause proved exceptionally difficult, as a result of Burmah's decreasing influence in the country, and the growing complexity of the government's overall pricing policies. It is only recent pressure from the World Bank and IMF which has succeeded in making an impact in this area.

Quite a large number of other oil companies have carried out some exploration work in Pakistan and then withdrawn, either due to problems in negotiating acceptable prices, or to technical difficulties in the fields or the quality of gas. Marathon, Sun Oil and Conoco all carried out unsuccessful exploration off shore, where there is a problem caused by high pressures. BP entered Pakistan in 1980, only to leave again in 1983. Gulf Oil left in 1982, after spending $50 million to no effect. Esso sold out of the Mari field, despite its firm sales contract with the fertilizer sector. Amoco and Shell also withdrew, although Shell showed renewed interest in the country in 1986.

10.6 Gas Policy, Contracts and Pricing

(a) *Contracts*. Traditionally contracts in Pakistan were designed on a cost-plus basis, to give companies a discounted cash flow of between 12 and 15 per cent on their investment. Companies complained that this gave them no incentive to explore for new fields. Even those fields that had been discovered tended to lie dormant, despite unsatisfied market demand for gas. For example Kandhkot was discovered in 1959, but only came into production at the end of 1986; Pirkoh was discovered by OGDC in 1977 but lay unused until 1984.

Companies also complained that gas prices were only fixed after a successful discovery was made, hence removing the companies' bargaining strength. Furthermore they were required to develop a field within six months, or else the government would step in and do so itself. It has been suggested that this led to companies not disclosing many finds. The prices for associated gas were especially low; and there was some feeling that the government did not always play fair on this issue, choosing to define gas as 'associated' even when there was a lot of gas and only a very small quantity of oil in a field.

In the face of these problems the government has tried to improve the returns available to oil companies. The pricing system has gradually been modified, to approach more nearly a market-based valuation of the different fuels, with the pricing principles being agreed on in advance of exploration. Meanwhile the exploration contracts have also taken on a new shape. Both the World Bank and the UNCTC have been involved in this process, with a new model contract being introduced in 1981. This is still criticized for not being flexible enough however, especially on its pricing conditions (see below).

The Dhodak field was offered for development bids in October 1986, under a risk service contract. This would require the operator to act as a contractor to OGDC. All development and operating costs would be reimbursed from a previously agreed share of gross revenues. In addition the contractor would be remunerated according to the results of the field. However, the company would have no ownership rights, and would be responsible for all foreign exchange risks.[13] The terms are not seen as being very favourable to private companies, and it has been suggested that they were framed in this way in order to discourage interest, since OGDC would in fact like to develop the field itself.

(*b*) *Pricing*. For many years Pakistan was renowned for having the lowest retail price for gas in the world. The government regulates prices at every stage, from production, through transmission and distribution, to the consumer. It followed an explicit policy of protecting the economy from the effects of high oil prices, and of encouraging high market penetration of gas through subsidized prices.

The well-head price payable to producers was traditionally based on actual costs plus a guaranteed rate of return varying between 12 and 15 per cent. At the time that Burmah drew up its original sales contract with the government this was generally considered to be an adequate return on investment; but since then companies' expecta-

---

[13] *Cedigaz*, 28 October 1986.

tions have changed considerably. The cost-plus formula failed to encourage the companies to make further investments, and as gas shortages began to be felt, the government attempted to improve returns to the companies. In 1982 a new formula was devised for Sui and Mari fields. For Sui the rate of return after tax was increased to 22.5 per cent. This meant an increase in well-head price from $0.08 to $0.22 per mcf. A 55 per cent increase in the well-head price also resulted for Mari. But this was still a modified cost-plus formula, and it still failed to generate the desired response from the companies. Finally in September 1985 a new system was introduced, whereby prices were set at 66 per cent of the international price for fuel oil, but with a negotiated discount for each producing group dependent on field location, exploration and development costs, market conditions, and so on. The exact terms of these agreements are not publicly disclosed.

Progress has been made under these new terms. Investors in the south of Pakistan have agreed to a sliding scale discount related to their volume of production, although those in the north are still negotiating. The government's intention is to stimulate accelerated exploration by private investors, and they have acknowledged the difficulties felt by companies. In particular they have recognized that following the 1986 fall in oil prices a well-head gas price of 66 per cent of fuel oil may be too low. But their plan is to review this formula in 1988, whereas quicker action would be much appreciated by the companies.

Consumer prices are also undergoing radical change. In the past they were determined primarily according to social and political objectives, and with the aim of encouraging wide-scale substitution of gas for imported fuels. Prices were therefore set at very low levels, and the policy was certainly successful in inducing rapid growth in the number of gas consumers. Faced with gas shortages since the beginning of the 1980s however the government has reformed its pricing strategy.

First, incentives for high consumption have been abolished. A flat rate has been introduced for most customers; and in the residential sector rates increase with the level of consumption. Secondly, under the terms of a World Bank Structural Adjustment Loan in 1981 the government agreed to gradually increase the average gas price, to reach two-thirds of the border price for fuel oil by 1988. This has led to dramatic price increases, averaging 140 per cent between 1975 and 1982. Coupled with the fall in international fuel oil prices, it is likely that the 1988 target will indeed be reached. These increases have been applied differentially in the various market sectors, bearing in mind especially the political and human impact of sudden increases on domestic consumers. The pattern of increases is given in Table 10.2.

Table 10.2: Consumer Prices for Gas in Pakistan. 1975–84.
Rupees per Million Cubic Feet.

Date	Domestic	Commercial	Cement	Fertilizer	Power	General Industries
1.1.75	6.40	10.32	4.7/3.2	5.68	4.7/3.2	5.6/5.3
1.2.75	9.60	13.76	6.4/4.3	5.68	6.3/4.3	7.5/7.1
28.6.79	9.64/12	18.84	9.5/7.0	7.10	9.5/7.0	11/10.5
1.1.82	14.0/21	22.00	13.00	11.85	12.00	13.00
9.1.83	16.0/21	25.34	16.34	15.19	15.34	16.34
11.6.83	16.0/24	29.63	20.63	17.63	19.63	20.63
14.6.84	18.0/?	33.94	24.94	21.94	23.94	24.94

Source: M. Ahmad, *Gas Consumer Pricing in Pakistan*, 1985, p. 9.

10.7 Markets

The first gas production in Pakistan was in 1955, and rates of consumption grew very fast, reaching 12 bcf by 1957, 170 bcf twenty years after that, in 1977, and 381 bcf by 1986.[14] This rate of growth was achieved through the government policy of maintaining very low consumer prices for gas, in order to encourage consumption of indigenous rather than imported fuels. As in other developing countries the power sector was a very early bulk user of gas, and fertilizer plants were also an obvious outlet for a country with such intensive agriculture. Pakistan is unusual however in the extent to which both the residential market and the general industrial market have also been exploited. The distribution of demand by sector, and its growth over the last fifteen years is shown in Table 10.3. The number of customers in each sector does not of course reflect the total consumption in those sectors. In March 1985 there were 4,171 industrial customers, 31,266 commercial customers, and 919,966 residential customers.[15]

The success of the gas promotion campaign was not matched by the gas supplies however. Low well-head prices gave little incentive to companies to seek out and develop new gas fields, and for some time supplies have been insufficient to match customer demand. For example during the winter of 1985 net gas production of 940 mcf/d fell short

[14] World Bank, internal report; GDC, direct communication.
[15] Ahmad, 1985, p. 6.

of peak demand by 360 mcf/d.[16] Shortages are particularly acute during the winter months, when demand increases by as much as 50 per cent.[17] Gas supplies to some large consumers have been curtailed, either completely (government-owned cement plants), or partially (sugar and power plants), and medium-sized manufacturers have been subject to intermittent supplies, perhaps three days a week. Residential and commercial consumers have also been hit by drops in supply pressure. The allocation of scarce supplies is the responsibility of the government (Ministry of Petroleum and Natural Resources), hence the pattern of these cuts can be seen as reflecting government policy on priority market sectors for gas. In 1981 the government asked cement factories to substitute fuel oil for gas, and this is clearly reflected in Table 10.3. Plans for new gas-fired power stations were also forestalled, with preference being given to the fertilizer and residential markets. As in most countries, political pressure is strong to maintain supplies to residential consumers, regardless of whether this is the most economic sector in which to concentrate gas use. Supply curtailments also tend to affect Karachi and sometimes Lahore, rather than the administrative capital of Islamabad.

Table 10.3: Gas Consumption in Pakistan by Sector by Fiscal Year (July–June). 1972–86. Million Cubic Feet per Day.

	1972/3	1977/8	1980/81	1985/6	% Share 1985/6
Fertilizer	76	87	181	273	29.6
Power	119	165	232	283	30.7
General Industries	83	131	185	205	22.2
Residential	8	27	49	116	12.6
Commercial	6	15	21	27	2.9
Cement	57	66	71	19	2.0
	349	491	739	923	100.0

Sources: World Bank, internal report; GDC.

(a) *Power.* The power sector is a major consumer of gas in Pakistan, as in other developing countries, accounting for nearly a third of all gas used in 1985/6. Gas use in the sector was developed very early: already by 1965 as much as 85 per cent of all thermal electricity generated, 1,814 GWh, was provided by gas. (Meanwhile hydro provided 1,362 GWh.) By 1979 gas had taken over 98 per cent of

[16] World Bank, internal report.
[17] Ahmad, 1985, p. 7.

thermal power production.[18] But the continuing fast growth in consumption of electricity and shortfalls in supply of gas have meant that the share of gas has gradually fallen since that high spot, with fuel oil providing the difference. By 1985 oil supplied 15 per cent of all power generated, gas 33 per cent and hydro 50 per cent.[19]

(*b*) *Industry*. As noted above industry accounts for a high level of gas consumption. This ranges from bulk consumers such as cement factories, right down to small workshops. Government policy has been to curtail supplies to the bulk consumers in times of gas shortage, presumably because switching to alternative fuels such as oil is easier for large concerns.

(*c*) *Residential and Commercial*. The cold winters in northern Pakistan, coupled with the favourable tariffs, explain the unusually high penetration of gas in this sector. Official figures refer to nearly 1 million households connected to the network (out of a possible 14 million); given the numbers of illegal connections, via rubber tubing running between houses, the actual number is probably considerably larger than that quoted in official documents. It is worth noting that the price of gas for residential consumers is much lower than the price of either LPG or kerosine. In March 1985 LPG cost 76 rupees and kerosine 86 rupees per 975,000 btu, whereas gas cost just 18 rupees for domestic users.[20] Consequently LPG tends to be mainly used in far-flung hilly areas, where there is no gas distribution network. There is however a shortage of LPG, and the government is actively discouraging its use in the transport sector, in order to conserve supplies for household use in remote areas.

(*d*) *Transport*. Given the overall shortage of gas to supply existing demand there seems little justification in introducing an innovative and expensive programme of converting vehicles to run on gas rather than petrol or diesel. There is however a minor CNG scheme in Karachi.

(*e*) *Fertilizers and Petrochemicals*. In fiscal year 1985/6 the fertilizer industry represented nearly 30 per cent of total gas demand. The gas from the Mari field has always been used primarily in three nearby fertilizer factories, Pak Saudi, Exxon and Fauji. However consumption of fertilizers in Pakistan is very high: it has the fourth-largest

---

[18] Government of Pakistan, 1983, p. 55.
[19] GDC, direct communication.
[20] Ahmad, 1985, p. 11.

irrigated area in the world, after China, India and the USA. In order to meet projected demand the Central Board of Revenue announced in October 1986 a five-year tax holiday to any new fertilizer or petrochemical plants, providing they were built within the next three years.[21] Given the overall shortage of gas in Pakistan, and low forecasted international prices for fertilizers for the foreseeable future, this seems to be a recommendation out of character with the main direction of government policy.

10.8 Finance

Financing the gas industry in Pakistan has been a constant problem for the government. The oil companies have been reluctant to commit large equity sums to projects, given the low levels of return. Meanwhile OGDC has had access to very restricted funds. The delayed development of fields such as Dhodak can be explained purely in financial terms (as mediated by the gas price). When it was first found in 1976 it was tipped for priority development; OGDC quickly initiated development wells, and by the second half of 1977 the government was anticipating production within two years. Schedules continually slipped however, as OGDC struggled to find the $70 million required for the first phase, and $200 million for the second phase.

Repeated experiences such as this have led to a lively debate within Pakistan over the advantages of private as against public investment. Should public funds carry the risk of exploration? What proportion of the national budget should be devoted to this sector? Local companies interested in investing capital were being excluded, which made little sense. The exclusive rights of OGDC have now been modified to allow the participation of private companies in exploration. Meanwhile funds available to OGDC have been given a boost: in May 1986 the government approved five projects worth nearly $104 million altogether.[22]

Much of the field development has been financed through international aid agencies. For example the World Bank, the Asian Development Bank and the Islamic Development Bank have all been involved in projects. The IFC has also invested equity in several schemes, including the Mari and Kandhkot fields. A $2 million syndicated loan was also extended towards the latter project, from the National Westminster, Morgan Guaranty and Standard Chartered banks.[23]

---

[21] *World Gas Report*, 6 October 1986.
[22] *World Gas Report*, 19 May 1986.
[23] *Oil & Gas Journal*, 31 December 1985.

Sources

Ahmad, M., *Gas Consumer Pricing in Pakistan*, paper presented to the Round Table Programme of the World Bank Energy Department, March 25–26, 1985.

GDC Inc, unpublished study.

Government of Pakistan, Ministry of Petroleum and Natural Resources, *Energy Year Book*, 1983.

International Labour Office, *Socio-Economic Aspects of Oil and Gas Development in Pakistan*, 1985.

World Bank, various unpublished papers.

In addition to these sources, and those cited in the footnotes, valuable information was gained from interviews as follows: British Gas, Mr Michael Cooke, Dr Aman Khan, Occidental, Mr Francisco Parra, Shell and the World Bank.

APPENDIX TO CHAPTER 10

Background Data

Population[a]: 96.2 million
GNP per capita[a]: $380
Average annual growth in GNP per capita, 1965–85[a]: 2.6 per cent
GDP[a]: $28,240 million

External public debt[a]: $10,707 million
External public debt as percentage of GNP[a]: 31.7

Total energy consumption[b]: 33,630 mtoe
   Commercial: 50 per cent   Non-commercial: 50 per cent
Commercial energy consumption per capita[a]: 139 kgoe
Commercial energy imports[b]: 6.281 mtoe
Commercial energy exports[b]: 0.304 mtoe

Proven Reserves and Production[c]

	Gas	Oil	Coal	Hydro
Proven Reserves	16 tcf	139 mbls	508 mt	19,600 MW
Production	38 bcf	14 mbls	2.2 mt	13,797 GWh

Consumption by Sector: Thousand Tons of Oil Equivalent[d]

	Gas	Oil Products	Coal	Total	% Share
Power	2,148	977	11	3,136	25
Industry	1,922	921	961	3,804	30
Commerce	232	19	–	251	2
Transport	–	3,386	–	3,386	26
Households	995	882	6	1,883	15
Agriculture	–	229	–	229	2
Total	5,297	6,414	978	12,689	
% Share	41	51	8		100

Power Sector[e]

Installed capacity: 6,300 MW

Electricity generation (%): Hydro=50, Oil=15, Coal=0.2, Gas=33, Nuclear=2

Sources

(a) World Bank, *World Development Report*, 1987, pp. 202–237. Figures refer to 1985.
(b) GDC Inc, direct communication.
(c) GDC Inc, with the exception of hydro data, which were supplied by the World Bank. All figures refer to 1986.
(d) GDC Inc. Figures refer to 1986. Data supplied by the World Bank give gas consumption by industry as 4.082 mtoe.
(e) GDC Inc. Figures refer to 1986. Data supplied by the World Bank give the following breakdown for electricity generation (%): Hydro=45, Oil=13, Coal=1, Gas=38, Nuclear=3.

11 TANZANIA

11.1 Country Summary

Tanzania is a good example of a low-income developing country, where at first glance it seems that exploiting the country's gas could make a vital difference to the economy. One of the government's major problems is finding hard currency; and oil imports currently use up more than 40 per cent of available foreign exchange.[1] A substitution programme therefore seems to be the obvious solution.

A closer look at the situation reveals problems however. The major difficulty is that the total potential market is barely large enough to justify investment in the infrastructure (the location of the gas reserves, off shore and distant from the major city exacerbates this problem). The limited market size is partly a function of the country's low level of industrialization, and partly a result of the large hydro potential in Tanzania. While gas could make a valuable contribution in certain market sectors there is no overwhelming case in its favour. A gas utilization study, sponsored by the World Bank, is currently assessing the economics of using gas in those markets which do exist.

In order to supply gas to such a small market the cost of at least some of the infrastructure must be covered elsewhere. One way of doing this is through an export project – since there is not enough gas for an LNG scheme this means either methanol or fertilizer. In theory such a project could in itself cover the costs of field development, a gas-gathering system and onshore pipeline. An export scheme would also bring the major advantage of actually earning hard currency – an even more tangible benefit than substitution of indigenous for imported energy. Great care is needed when committing the reserves to an export project however. The proven gas reserves of 750 bcf are large in relation to Tanzania's energy consumption; but they cannot support more than one relatively small export project. The market risk associated with such a project is also very high for a low-income country. A collapse in the price of the commodity produced could negate the whole purpose of carrying out the project, which is to earn

---

[1] Ministry of Energy and Minerals, direct communication.

a return. The government's direct liabilities in case of failure may be limited through careful negotiation of terms; but using up an irreplaceable natural resource calls for a clear positive prognosis, not just the covering of risks.

The implementation of such a project could also prove hazardous for other reasons. Any such scheme requires large capital investments, which are not easy to secure for a poor country such as Tanzania. The time-lag in building up sufficient resources of equity and debt means that feasibility studies for the projects themselves become outdated. For example, finance has now been found to build a fertilizer plant, but seven years after the time at which the fertilizer market was originally judged buoyant enough to justify such a scheme. Market prospects have waxed and waned since that time. While the most recent studies still support the project, there is still considerable doubt in some quarters over the wisdom of the scheme. There is a danger that availability of funds will of itself be used to justify construction of the plant. Investors and lenders often use very different criteria in assessing projects to those which an objective outsider would employ, even when there is no government guarantee to cover their risks. The interest of construction companies in maintaining business is well known; export credit agencies have similar self-interested goals; and even aid agencies operate under pressure to disburse available funds. After many years of planning a project vested interests also come into play. All those concerned become anxious to see a concrete result for all the planning and money that has already been invested in it.

Tanzania is thus a border-line country, where special care needs to be taken in assessing both the prevailing international market for gas products and the prospects for market development within the country, before launching a gas utilization project. There is always a threshold below which the necessary capital investment cannot be justified on commercial grounds. It may even be that aid finance could be more profitably employed elsewhere in the economy. If a negative decision is reached however, at least the gas will still be in the ground as a resource for future generations.

11.2 The Energy Balance in Tanzania

The quantities of gas available are large relative to commercial energy consumption in Tanzania. In 1985 only 0.6 mtoe of commercial energy were consumed in the country;[2] and proven reserves at Songo-Songo alone amount to 17 mtoe. At a deliverability of 70 mcf/d,

---

[2] *Ibid.*

Songo-Songo could provide for Tanzania's total petroleum requirements for the next twenty-eight years, if consumption levels do not increase.

Tanzania is a very poor country, with per capita GNP of $290 in 1985, and an average growth rate of less than 0.5 per cent between 1965 and 1985.[3] The country's economic problems have been exacerbated by the need to import as much as 90 per cent of commercial energy,[4] in the form of crude oil and oil products. These imports have swallowed up a large proportion of Tanzania's 'free' foreign exchange. Matters have at times deteriorated to the point where delivery of a much-needed cargo of oil has been delayed while the hard currency to pay for it is found. The net result has been frequent shortages of fuel and rationing. Industries are forced into short working, and the economy deteriorates even further in a vicious spiral. Under these circumstances the major potential benefit from finding natural gas is seen in terms of the hard currency it can generate. This may be either through an export project, or through substituting gas for imported oil. The latter has proved difficult, since infrastructure costs are high, only a limited bulk market in the power sector exists, and in any case more than half of the imported oil is used in the transport sector. The possibilities are still under discussion; but the emphasis has fallen primarily on the export market, as first priority. The impetus behind the planned fertilizer plant has come from the country's urgent need for hard currency to meet existing and future commitments as part of the current economic recovery programme.

11.3 Gas Reserves, Production and Infrastructure

(a) *Reserves*. Gas was first discovered in Tanzania in 1974, off shore at Songo-Songo, and the availability of commercial quantities was established in 1977. Further reserves were found in 1982, at Mnazi Bay, also off shore, although considerably more drilling and appraisal has been carried out at Songo-Songo than at Mnazi Bay. Both fields are of non-associated gas, and are almost entirely methane. Both sites are distant from Dar es Salaam: 137 miles and 288 miles respectively. The size of the reserves is given in Table 11.1.

No gas has yet been produced from either field, although numerous utilization studies have been carried out, first by AGIP who discovered both fields, and later for the government after AGIP had relinquished ownership. There are hopes that a major export fertilizer

---

[3] World Bank, *World Development Report*, 1987.
[4] World Bank, internal document.

Table 11.1: Tanzanian Gas Reserves.

	Songo-Songo	Mnazi Bay	Total
Proved	726	23	749
Probable	157	586	743
Possible	223	–	223
Total	1,106	609	1,715

Source: World Bank, internal document.

project will start in the near future, possibly followed by some domestic use of the fuel.

The major problems in developing domestic gas use have been the costs of bringing the gas on shore, the distance of reserves from any potential market, and the absence of a clearly defined market of substantial size. If the fertilizer project should come to fruition the costs of a gas-gathering system will not fall exclusively on the domestic scheme however. Detailed market studies were carried out for the power sector in 1985, and are now being completed for the industrial sector.

(*b*) *Infrastructure*. Bulk use of gas will require the construction of a pipeline to Dar es Salaam, since half of all commercial energy is consumed in and around the capital city. This is a distance of 137 miles, and crosses difficult terrain, including the Rufiji river basin. This area is very swampy; in one recent rainy season the river moved four miles across the delta. It would be unrealistic to expect either the government or private capital to provide finance for such a large investment, given the perilous state of the country's financial position, and uncertainty of the market. Aid would therefore be required.

An alternative possibility has also been looked at, but unfortunately has been found unviable. This was to transport gas in the form of CNG in 2-ton tanks on barges, which could have sailed from Songo-Songo direct to Dar es Salaam. This is technically feasible, and could provide fuel for use in vehicles and perhaps for commercial premises, thus opening up and establishing a small market without making such a large investment in infrastructure. However studies have found that gas delivery costs to consumers would range up to $30 or more per mcf of gas, and are thus prohibitive.

11.4 The Government Sector

Overall policy-making powers lie with the Ministry for Energy and Minerals. TPDC (the Tanzania Petroleum Development Corporation) is involved on the basis of production-sharing agreements with exploration programmes, and will have operational responsibility for any gas utilization.

(*a*) *Ministry for Energy and Minerals*. The Ministry determines overall energy policy, and since the early 1980s has placed gas utilization high on its list of priorities. Various studies of the potential market for Songo-Songo gas have been carried out for the Ministry, including one by Norsk Hydro and one by IGDC.

Decision-making is relatively centralized. Ultimate authority rests with the Minister, who was also Chairman of TPDC and of Tanesco (Tanzania Electric Supply Company) until 1986. The Board Chairmen and Chief Executives still consult the Minister on all important policy issues, so there are no apparent conflicts of interest between these institutions, nor any complex structure of committees to delay decision-making. Under these circumstances the attitude of the Minister himself towards gas is very important. In fact he has taken a great interest in promoting the use of gas. The decision to plan for a world-scale fertilizer plant was taken at an early stage, and has been pursued with persistence. Progress has been slow, but this has been due to the inherent problems in raising finance, and not to any lack of enthusiasm. The present Minister's former position as Minister of Finance for the now defunct East African Community puts him in a good position *vis-à-vis* the Ministry of Finance; this must have been of use when negotiating tax rates in exploration contracts, or arguing for the fertilizer plant to be the first enclave project in the country. Other senior executives in the Ministry include the Energy Commissioner, who is a former university professor, and an expatriate adviser on energy matters. There have also been close links with the Commonwealth Secretariat, which has provided much advice and direct help in drawing up the Petroleum Act and in negotiating contracts.

(b) *Tanzania Petroleum Development Corporation (TPDC).* The company was established in 1969, as the local partner for production-sharing and joint-venture arrangements with oil companies exploring for oil and gas. TPDC also controls the sale of all imported oil and oil products to the distributors. As a result it is the richest company in Tanzania, and has the ability to provide most local financing required in exploration or, for example, in the fertilizer project. As a government-controlled body it can also boost these funds if a special need is foreseen; although this capacity can be limited by outside events. For example when the price of oil fell dramatically in 1986 TPDC maintained the same selling price to distributors, with the intention of amassing the difference as extra cash in anticipation of the need for equity in the Kilamco project. However, 50 per cent depreciation of the Tanzanian shilling against the dollar in the first half of 1986 more than offset the savings. This was further exacerbated by the fact that since 1982 TPDC had imported significant quantities of oil on deferred payment terms.

TPDC is well supplied with qualified engineers and geologists, a number of whom have been trained by Statoil and at Norwegian universities through a NORAD aid programme. Inevitably, however, they are lacking in hands-on experience in production. When Songo-Songo is developed, it is anticipated that a service contract with a foreign company will be needed at first, while local staff are trained on the job. There is also a problem of lack of incentives for staff. Working for a foreign company is certainly seen as more desirable than working for TPDC, thanks mainly to the numerous fringe benefits that are available.

A special company, Gasco, has been incorporated in anticipation of future developments in gas, but as yet has no personnel. For the time being a Gasco/Kilamco unit within TPDC is responsible for monitoring the progress of gas planning in Tanzania. It is expected that this unit will eventually provide the core staff for Gasco.

11.5 The Private Sector

Only two companies are currently commercially active in Tanzania, Shell and Amoco. Shell has a concession area south of the Ruaha/ Rufiji rivers in the Selous game reserve, and Amoco is exploring in the south-west of the country, in the Rukwa and Lake Tanganyika basins. Both companies are looking for oil, not gas. Neither exploration contract enters into details of conditions under which gas finds may be exploited.

Over the last few decades a number of companies have been active in Tanzania, but most left the country when they failed to find any oil. Shell and BP explored off shore between 1950 and 1961, but left empty-handed (Shell and Esso returning again in 1981). IEDC, Elf-Aquitaine and KUFPEC also carried out exploration programmes, but without making a discovery. Both gas finds were made by AGIP with Amoco. Songo-Songo was discovered in 1974, and Mnazi Bay in 1982. AGIP is said originally to have been quite enthusiastic about the Songo-Songo gas find, and considered especially its use in a methanol plant, to meet increased international demand in the transport fuel sector. Market studies discouraged this path however, and they relinquished the field to TPDC in 1975. TPDC then drilled two wells, in 1976 and 1977, when gas was proven in commercial quantities. A full appraisal of Songo-Songo was subsequently carried out for TPDC under a World Bank Assistance Programme; Mnazi Bay has never been fully assessed.

Two other companies have been involved in exploration in Tanzania, but under aid programmes. These are Statoil and Petro-Canada International Assistance Corporation. Statoil has also carried out extensive training of geologists through the Norwegian Petroleum Directorate, both in Norway and Tanzania – to the extent that they feel TPDC is now well equipped with personnel. But in terms of exploration these companies have been occupied with laying the groundwork to attract commercial interest from others. They carry out preliminary surveys, to define areas which might be of interest in future licensing rounds. While this is a very useful function, it also puts the companies concerned in a rather ambiguous position. As commercial companies themselves, or with links to a commercial company, they gain privileged knowledge. And they are in a difficult dilemma when drilling exploratory wells; for if they choose a promising area then other companies are likely to be interested themselves and feel that the aid donor is gaining unfair commercial advantage over its competitors. This can lead to some strain between the parties involved.

11.6 Gas Policy and Contracts

All exploration contracts in Tanzania are backed by the 1980 Petroleum (Exploration and Production) Act. This was drafted in collaboration with the Commonwealth Secretariat, to provide a stable legal framework within which oil companies could comfortably operate. For example it contains a provision whereby the Minister cannot refuse to renew a licence except for specified reasons laid down in the Act. It succeeded in dispelling the unease which resulted from an earlier spate of nationalizations, but at the same time it protects the interests of the country. For example, a number of less scrupulous small independents and individual prospectors who sensed the possibility of windfall profits were soon deterred by its requirements.

Within this general framework, the actual production-sharing agreements tend to contain no special provisions for natural gas. In this they are similar to those in many other countries. This has not been a cause of special concern to companies, however, since they have an overriding pessimism about the difficulty of developing a market for gas in Tanzania. The market problems will be elaborated on below. But suffice it to say, that companies are not over-worried about the intricacies of taxation terms, or production shares, when there is in any case no obvious outlet for the product.

The only terms specified in the standard contract are that an additional five years (to the time allowed for oil) will be given for assessment of the field prior to application for a Development Licence; that gas should be valued as far as possible in relation to alternative fuels; and that the shares accruing to the contractor and to TPDC should be determined in the same way as for oil (these being dependent on average daily production rates and the extent of TPDC's participation in joint operations). Thus, the gas clause leaves most details to be negotiated 'in good faith' when the occasion arises. In the case of Tanzania, this should perhaps be interpreted as neither side realistically anticipating a company showing serious interest in developing a gas find. Nevertheless, it is true to say that the companies do anticipate fair treatment from the government, and to that extent the phrase 'in good faith' is meaningful.

The only company to have shown interest in defining the gas clause more closely was Elf-Aquitaine, which spent a long time drawing up a detailed gas clause when seeking a concession near to Songo-Songo. This clause was drawn up in conjunction with the Commonwealth Secretariat. It contained much greater detail on the periods allowed for field appraisal and commercial assessment. It defined the rights of TPDC and of the company, and provided that the company would be compensated if they chose not to develop a field which TPDC did develop for a specific market. However, in the event Elf never carried out exploration work under this contract; it suddenly withdrew from negotiations and instead farmed into an area further north, with IEDC.

11.7 Markets

The prospects for developing the internal gas market in Tanzania will be greatly affected by the construction of the fertilizer plant at Kilwa, should this go ahead. For this will bring the gas on shore from the Songo-Songo field, making it more accessible to markets; but it will also require about half of proven reserves of gas from the field, thus limiting the scope of market development. A new utilization study is now being carried out by GDC, and will be completed in 1987. Government planners are therefore awaiting the results of this study before drawing up firm policies.

(*a*) *Power.* Gas could potentially find a bulk market outlet in the power sector in Tanzania; and this could provide the economic justification for building a gas pipeline to Dar es Salaam, which could then also serve the relatively small industrial market. However the decision to follow this course will depend on a complex weighing of the advantages of gas-fired power stations as against hydroelectricity. There is not the clear case for using gas that there is in countries where it would substitute for imported oil, where there is a constant drain on foreign exchange. In fact comparatively little foreign exchange is used in purchasing fuel for the power sector in Tanzania, since as much as 87 per cent of all electricity was hydro-generated in 1985.[5] The balance is mostly provided by small diesel stations in remote parts of

---

[5] Ministry of Energy and Minerals, direct communication.

the country, where gas substitution is not a practical proposition.

Furthermore, electricity accounts for only a relatively small proportion of commercial energy use in Tanzania, and demand is not expected to grow very fast, unless there is a dramatic turnaround in the economy. Only 3 per cent of the population have access to electricity.[6] Industry is the largest consumer of power, but even this sector does not include many energy-intensive processes. The fifteen most energy-intensive concerns spend on average only about 5 per cent of their costs on electricity.[7]

It can also be disputed whether finding a bulk market for gas in the power sector is in fact desirable in Tanzania, given the relatively limited size of proven gas reserves. Every 100 MW of added power capacity requires approximately 0.2 tcf of recoverable gas reserves, and the Kilamco project will require approximately half of current proven reserves. Under these circumstances it is arguable that industry should be given priority use of reserves, where gas will be substituting directly for imported fuel oil.

The case for hydroelectricity is strong. It already provides nearly three-quarters of all electricity in Tanzania; but there is still an enormous amount of unused hydro potential – an estimated total capacity of 20,000 GWh per annum, compared to the 695 GWh actually generated in 1983.[8] Future growth in demand for power could therefore be satisfied quite easily from this source, without the need for expensive imports of fuel. Furthermore this latent hydro capacity includes a large number of small sites which could be developed in remote areas, should demand for power justify this. Hydroelectricity is also of course a renewable source of energy; and in the past has been attractive to aid agencies, because large dams provide a focal point for demonstrating the results of aid programmes.

Despite all these advantages of hydropower, a strong case can also be made for the use of natural gas to complement it. While final decisions have not yet been taken in Tanzania, it seems likely that if gas is to be used at all for the internal market then it will penetrate this sector at least to some extent. There are several reasons for this, in addition to the usual requirement of creating sufficient demand to justify a transmission pipeline. First, there is a danger that the power system may become over-dependent on hydroelectricity, putting it at risk in times of drought. Gas could provide some measure of diversification. Secondly, the construction of gas combined-cycle power sta-

---

[6] World Bank, internal document.
[7] Tanesco advertisement in *Daily News*, 3 June 1986.
[8] World Bank, internal document.

tions to meet future increments in demand will be less expensive than hydro stations. While the latter do not require foreign exchange for fuel, they do require very large amounts of initial capital; gas combined-cycle power stations are considerably cheaper to build, per MW of electricity produced. Thirdly, they can be constructed with only a short lead time, and typically provide much smaller capacities. This means that their construction can be much more closely matched to actual demand, leading to higher utilization of installed capacity. This in turn makes them less costly in absolute terms than the hydro plants, which need to be planned much earlier in advance of demand, and in much larger increments.

(*b*) *Industry*. Boiler and furnace applications can be identified as the prime target for gas substitution. This sector is a major consumer of oil and oil products, swallowing large amounts of hard currency. The Wazo Hill cement plant on its own uses 10 per cent of all the fuel oil consumed in Tanzania;[9] fuel constitutes between 40 and 50 per cent of its total manufacturing costs. Other potential consumers tend to be quite small industries, such as glass, textiles, paper and brewing. Between them however they consume 27 million litres of fuel oil a year, equivalent to approximately 1.5 bcf of gas.[10] Although the industries are small, they have the advantage that they are relatively centralized. Over half of all the commercial energy consumed in Tanzania is used in Dar es Salaam, and most industries are on the periphery of the city, with a concentration near the airport. Provision of distribution lines would therefore be relatively straightforward. The Wazo Hill cement plant is about twelve miles north of the city, but as the largest consumer of hydrocarbons in the country it constitutes a special case.

(*c*) *Residential and Commercial*. The residential sector depends overwhelmingly on firewood in rural areas, with some charcoal consumption in urban areas. Together these provide 99 per cent of all energy consumed in the sector.[11] Only about 8 per cent of the population live in cities (nearly half of these in Dar es Salaam), so there is little prospect of the residential sector providing a large market for natural gas. There is some scope for use of natural gas in hotels and offices, especially for air-conditioning. But once again this would be competing primarily with electricity provided by renewable resources; it would not therefore constitute a large saving of foreign currency.

[9] Ministry of Energy and Minerals, direct communication.
[10] IGDC, *Gas Utilization Study*, 1983.
[11] World Bank, internal document.

(*d*) *Transport.* At first sight this appears to be a hopeful market prospect, in that transport absorbs 54 per cent of the petroleum products consumed in the country.[12] Since all oil is imported it requires large amounts of hard currency. Furthermore, lack of hard currency means that there are frequent shortages of fuel, rationing, and a flourishing black market. Some interest is being shown in the possibility of using CNG in the transport sector, and this will be covered by the new utilization study.

Over half of all vehicles are owned by the government (about 25,000 vehicles). This would make it quite easy logistically for the government to introduce a CNG conversion programme. There is a problem however, in that the condition of vehicles tends to be poor, as a result of use on bad roads. The break-even point for conversion is about four years; if it is doubtful whether a vehicle will still be on the road that far ahead, then neither a private individual nor a government body is likely to want to invest in the conversion.

A conversion programme requires quite large amounts of foreign currency in the first instance, even though it may save foreign exchange in the long run. In the case of Tanzania it would prove especially difficult to raise this funding. For example, the LPG installations that already exist are in a chronic state of disrepair, almost to a dangerous extent, due to lack of foreign exchange for buying spare parts. This has reached the stage where AGIP has decided to donate the necessary spare parts, rather than risk major breakdown of the system and increasingly frequent accidents.

Despite these difficulties it is possible that CNG conversion should be recommended for carefully selected fleets of vehicles, perhaps buses or delivery trucks. But such a programme would need very careful targeting, and should not be applied in a blanket manner.

(*e*) *Kilwa Ammonia Company (Kilamco).* Plans for Songo-Songo gas have always given high priority to export projects, due to the limited internal market, and to the country's urgent need to earn foreign exchange. The size of the gas reserves and difficulty in raising finance meant that it was unrealistic to envisage more than one export project, at least in the first instance. Early utilization studies favoured a fertilizer project over methanol, based on market forecasts made in the early 1980s, and the fact that there was not enough gas for an LNG project. The consensus remains that prospects for new methanol plants are bleak.

Planning for the fertilizer project started in 1979, when Songo-

---

[12] *Ibid.*

Songo was relinquished by AGIP and came into the hands of TPDC. It has still not reached fruition however, largely due to delays in raising the necessary finance. Since such delays can often be a problem in gas developments, the history of the project is given in full. The upturn in the fertilizer market which was forecast for the mid-1980s did not occur, but new market studies carried out in 1987 by a consultant and by the World Bank (the latter using more pessimistic assumptions) both suggest that the project remains viable. Nevertheless, final commitment of funds from the financing bodies still awaits the World Bank's formal assessment report on the project.

The plan is to build an ammonia/urea plant at Kilwa, which is located on the coast close to Songo-Songo. It is to produce 1,560 t/d of ammonia, and 1,750 t/d of urea, 95 per cent of which will be exported. The plant will require construction of a gas-gathering system and onshore pipeline, as well as a deep-water harbour for the export trade. Consequently the total cost is extremely high: originally put at about $650 million, but now reduced to $425 million. From the beginning it was resolved that the project would only go ahead if this finance could be raised without government guarantees. Apart from practical difficulties in offering such a guarantee, the attitude of the government has been that if funds were committed on such terms this in itself would provide reassurance that the project was indeed sound. The availability of funds has therefore come to be treated as a test of the soundness of the project. Under these circumstances a great burden is placed on the terms of financing. The government needs to be very confident that lenders, and equity holders, will only receive a just return on their money, and that TPDC and the Tanzanian nation do not carry more than their share of the risk. Great care has indeed been taken in drawing up the terms, especially in regard to equity holders.

The original scheme, initiated in 1981, formed a joint-venture company between TPDC (74 per cent of equity) and the US firm Agrico (26 per cent). Agrico was to be responsible for three roles: (a) to take the technical lead in the project; (b) to manage and operate the plant; and (c) to market and sell the products on a commission basis (not take-or-pay). The plant was to be built in Sweden and floated out to Tanzania – the first time that a fertilizer plant would have been constructed in this way. There were a number of problems with the scheme. First, costs were extremely high, (estimated at $650 million), especially when compared to the costs of competing Middle Eastern plants. The Swedes were offering finance on aid terms, but this proved to be somewhat tentative, since they were also negotiating similar deals with Bangladesh and with Thailand, and could not afford all three schemes. Secondly the fact that the technique was experimental added unwanted risk. Thirdly, the Swedish construction company was unwilling to either give a completion guarantee, or to take on project management. The latter was essential, with components being assembled from around the world to various specifications. Fourthly there was a potential political problem with carrying out all the work in a developed country; even the construction would have provided no local employment. Negotiations continued for several years, attempting to solve some of these problems. But ultimately the major problem was the cost of the scheme. The IFC, for example, was very critical. Under these circumstances it was very difficult to raise sufficient capital.

A turning-point came in November 1984 when Kellogg joined the scheme, taking half of Agrico's equity interest. Kellogg had wide experience of the fertilizer business, having constructed a number of ammonia/urea plants around the world. They therefore had an off-the-shelf design, which they could offer as a turnkey project at a fixed price. Construction would also take place on site. Kellogg would share technical responsibility with Agrico, but would take on the role of managing and operating the plant once it was built. This would leave Agrico with responsibility for selling the products, which reflected their particular expertise. Most importantly, the new scheme would cost considerably less, at $425 million. In December 1986 Agrico withdrew from the venture, and has since been replaced by Trans-ammonia, on identical terms.

The debt/equity ratio is to be 3:1. In other words, of the total cost, $319 million is to be debt, and $106 million equity. Kellogg and Transammonia will each hold 13 per cent of the equity. TPDC's share has been reduced to 51 per cent for the duration of long-term debt, in order to satisfy the rules of the Paris Club. The IFC and Commonwealth Development Corporation will now take up the remainder. TPDC is able to supply the local currency component with little difficulty (see above). It can also raise some of the necessary dollars through signature fees on oil exploration contracts. But it still needs to raise almost $30 million from banks, and this is yet to be achieved. The debt share of the finance now seems to be available. Sources include Belgium, China, OPIC, Sweden, the UK and Yugoslavia. (Earlier offers from Austria and Italy were withdrawn when construction was no longer to be at Swedyards, since the technologies available through these countries' export credits were no longer appropriate.) The only remaining problem on debt financing is that each party retains the right, up to final signature, to withdraw its offer if it so decides.

The relationship between the equity partners has been designed to share the market risk equitably between TPDC, Transammonia and Kellogg. The features of the agreement are as follows:

(a) It will be a turnkey project, at a fixed price for lump-sum payment, with completion guarantees.
(b) Income from sale of the ammonia and urea will go into an escrow account in London, to service creditors as first priority. (This is the first time such an arrangement has been allowed in Tanzania, and agreement to it represents a considerable political achievement.)
(c) A debt service reserve is to be established. No profits will be distributed until one year's reserve debt servicing has been built up in foreign exchange.
(d) Until this debt service reserve has been accumulated each equity holder is party to a subordination scheme. That is, they defer their own return, on a *pari passu* basis with the other equity holders. Specifically:
 1. TPDC will allow the gas price to fall as low as is necessary to amortize the soft loan for building the pipeline system from Songo-Songo to the plant site;
 2. Kellogg will subordinate its operating bonus;
 3. Transammonia will subordinate part of its marketing fee.
(e) The gas pricing formula allows that if world ammonia/urea prices fall to unexpectedly low levels, then the gas price may fall to as low as $0.15/mbtu (from about $0.75). This could allow the plant to continue in operation through a bad patch when less flexible competitors might be forced to close down. However there is also a provision for a gas bonus to be paid when an adequate return on equity has accrued to the shareholders.

Under this last clause a considerable risk is to be borne by TPDC. The major concern therefore remains, whether the fertilizer market is buoyant enough to support Kilamco. At the time of the IGDC study in 1983, the outlook for the fertilizer market appeared rosy, with a sharp price increase expected in the mid-1980s. Tanzania was judged to be in a favourable geographical position, since the major export markets are in India and the Far East, as well as in Africa itself, and a port at Kilwa would be in prime position to reach these markets. Efforts have been made to secure these markets; for instance China will accept repayment of its debt finance in kind rather than in cash, and neighbouring countries may well be able to allocate aid finance to purchase of Tanzanian fertilizers. But there are no firm take-or-pay agreements for the products of Kilamco. Under current pessimistic forecasts for the fertilizer market this remains a serious worry. It is unlikely that work will actually start on the plant until this can be dispelled.

11.8 Finance

The economy of Tanzania has been in crisis for some time. External public debt amounted to 48 per cent of GNP in 1985,[13] and economic growth has been very slow, or even negative in some years. As a result there have been severe restrictions on moving hard currency out of the country, and the local currency has been subject to rapid devaluation. (Oil companies involved in distribution of products in Tanzania have resorted to investing in land in order to maintain the value of their frozen funds.) It has therefore been a challenge to raise private capital for investment in the country.

This situation is especially relevant to any plans to develop gas utilization within the country. As has been mentioned the Ministry for Energy has been successful in persuading the government to allow Kilamco as an enclave project, thus attracting private financing. For domestic developments however the government must rely on aid finance, whether from the World Bank, other multilateral institutions, or individual governments.

The World Bank has already financed several gas programmes, including contributions towards the appraisal of the Songo-Songo field and the new utilization study. Other aid finance has come from the OPEC Fund, the European Investment Bank and ONGC of India, as well as the programmes financed by NORAD and PCIAC mentioned earlier.

Sources

In addition to the sources quoted in the footnotes, information in this chapter was obtained from interviews with executives from the following companies and organizations: AGIP, Amoco, Elf-Aquitaine, GDC Inc, Ministry of Energy and Minerals (Tanzania), Shell, Statoil, the Tanzanian Electric Supply Company, the Tanzania Petroleum Development Corporation and the World Bank.

---

[13] World Bank, *World Development Report*, 1987.

APPENDIX TO CHAPTER 11

Background Data

Population[a]: 22.2 million
GNP per capita[a]: $290
Average annual growth in GNP per capita, 1965–85: <0.05 per cent
GDP[a]: $5,600 million

External public debt[a]: $2,988 million
External public debt as percentage of GNP[a]: 48.6

Total energy consumption[b]: 9.562 mtoe
   Commercial: 7 per cent   Non-commercial: 93 per cent
Commercial energy imports[c]: 630,000 toe
Commercial energy exports: n/a

Proven Reserves and Production[b]

	Gas	Oil	Coal	Hydro
Proven Reserves	749 bcf	–	304 mt	9,500 MW
Production	–	–	4 ktoe	222 ktoe

Consumption by Sector: Thousand Tons of Oil Equivalent[c]

	Gas	Oil Products	Coal	Total	% Share
Power	–	27	–	27	4
Industry	–	130	3	133	21
Commerce	–	17	–	17	3
Transport	–	342	–	342	54
Households	–	53	–	53	8
Agriculture	–	31	–	31	5
Other	–	30	–	30	5
Total	–	630	3	633	
% Share	–	99.5	0.5		100

Power Sector[d]

Installed capacity: 370 MW
Electricity generation (%): Hydro=67, Oil=33, Coal=0, Gas=0

Sources

(a) World Bank, *World Development Report*, 1987, pp. 202–237. Figures refer to 1985.
(b) Ministry of Energy and Minerals, Tanzania, direct communication. Figures refer to 1985.
(c) World Bank, direct communication. Figures refer to 1985.
(d) World Bank, internal paper.

12 THAILAND

12.1 Country Summary

Substantial gas reserves were discovered in Thailand towards the end of the 1970s, and the government drew up a positive policy to use the gas to the country's benefit. It planned to establish first a bulk market in the power sector and cement industry, to be followed by the development of 'higher-value' uses in petrochemicals and fertilizers, using the gas as feedstock. It was successful in building up the bulk market; but as yet the ambitious Eastern Seaboard industrial development programme has not materialized, although a contract has recently been awarded to start work on a petrochemical complex. Even the success has not been without its problems however. First it had to contend with a surplus of fuel oil from Thai refineries, displaced in the power sector by gas. More recently, it has found it difficult to maintain the market share of gas in the face of relatively low oil prices. Wells have been closed in, and exploration has more or less ceased. It has also proved necessary to cut drastically both the well-head and selling price of gas, to the point where the gas producer claims to be barely breaking even.

Thailand is an interesting case to consider, because the focus is on political and economic issues. If a project can be shown to be profitable then both private and public sources are willing to invest in Thailand. Unusually for developing countries raising finance is not therefore a problem.

The relationship between the government and Unocal, the major gas-producing company in Thailand, provides an interesting case-study. The room for bargaining and negotiation between a single supplier and a single customer is very limited. This mutual dependence has been pointed out recently by the President of Unocal Thailand. The company is in a 'unique role in Thailand. Our company has never before in its 100 year history been in a situation where the success or the failure of its activities has such an impact on the host country. It is a responsibility we feel very keenly, and are always mindful of.'[1] Under current market conditions the government is in a

[1] *The Nation*, 24 July 1986.

favoured position in this relationship, because it can easily find cheap alternatives to natural gas. Unocal has essentially no room to man- oeuvre; moreover, the companies are suffering from greatly reduced revenues at the moment even from exportable crude oil. In the longer term Thailand will need to use its gas reserves, and for the time being is heavily dependent on Unocal as its main supplier. There is there- fore room for the balance of power to be reversed.

Whilst other gas fields have been discovered, the government has failed to reach production agreements with any other company (with the exception of purchase of associated gas from Shell). The reason for the failure of these other negotiations is explored in this chapter. A very important factor is the oil companies' perception that govern- ment decision-making is far too slow. These delays are due to the complexity of the issues involved, and to a political decision-making system that seeks to achieve consensus, balancing the interests of the many state institutions involved. Apart from procedural delays however, the Thais had good reasons not to hurry into agreements with more gas producers. Once the Unocal field was being exploited more than enough gas became available to meet demand through to the 1990s. Furthermore Texas Pacific's gas was of lower quality than Unocal's and located further off shore. Very recently negotiations with Exxon have taken a step forward, and prospects of production from their inland gas field now seem hopeful.

The major problem in Thailand has been the matching up of demand with supplies of gas. A sequence of unpredictable events, outside the participants' control, has undermined the creation of a secure market for natural gas. First, the gas producer failed to meet predicted output levels, as the size of reserves was dramatically down- graded. This forced waiting customers to revert to burning fuel oil despite their heavy capital investment in gas-burning equipment. This was not a good start, and left consumers with a legacy of mistrust and uncertainty around gas supplies. Nevertheless this problem was being overcome, when a dramatic fall in the price of oil in 1986 reduced the interest of customers in using gas. While the original take-or-pay agreement was still honoured, the producer's hopes of building up and increasing the market suffered a set-back. At the same time the government's reliance on increasing LPG supplies as a by-product of greater gas production was threatened, adding a further twist to the situation. Adjustment of relative energy price levels has prevented total disaster; but the situation remains tentative.

Both the government and oil companies found themselves trapped by schemes which came into existence under very different conditions. The most remarkable aspect of gas development in Thailand has been the way in which they have survived these problems which have not been of each other's making. Both sides have made compromises, and the government in particular has adjusted some of its policies (on pricing and on the ownership of pipelines for example). As a result gas use is still developing, and negotiations are still progressing, despite the difficulties which have arisen.

12.2 The Energy Balance in Thailand

The pattern of energy production and consumption in Thailand has changed considerably over the last five years, as a result of the policies pursued under the government's Fifth Plan (1982–6). In 1982 traditional fuels (wood, charcoal, etc.) accounted for 38 per cent of secondary energy demand, while 52 per cent of demand was for oil products, virtually all of which were imported. The main direction of energy policy has been to increase Thailand's self-reliance in energy, through increasing exploration for and exploitation of indigenous natural resources. By 1985 the share of traditional fuels had fallen to 30 per cent and imported fuels to 40 per cent; the remaining 30 per cent was supplied from indigenous sources.

Oil and gas production increased from minimal levels in 1980 to 35,000 b/d of oil and 327 mcf/d of gas in 1985. Lignite deposits in the north of the country at Mae Moh had also been developed, and hydro generating capacity increased from 1,317 MW in 1981 to 1,816 MW in 1985. Nevertheless, imports of crude oil and oil products in 1985 amounted to 67 million barrels, costing the country more than $2,000 million.[2] The Sixth Energy Plan therefore continues to stress the need to develop indigenous resources and to reduce imports. Gas is envisaged as playing a very important role in this scenario, with demand forecast to rise from 230 mcf/d in 1984 to 600 mcf/d in 1991, and potentially 1,100 mcf/d in 2001.[3]

12.3 Gas Reserves, Production and Infrastructure

(a) *Reserves and Production.* Proven remaining reserves of natural gas in Thailand as of 1985 amounted to 3,721 bcf, and were located in four areas. As can be seen from Table 12.1, the large gas deposits so far located have been off shore, in the Gulf of Thailand. There has been a move towards increased exploration on shore in recent years, especially in the central and northern regions of the country. Hopes for this

---

[2] Hongladaromp, 1986, p. 1.
[3] Most of the information in this section is taken from NESDB.

Table 12.1: Gas Reserves in Thailand.

Oil/Condensate (Million Barrels)	Gas bcf	Operator	Location
48.3	1,517	Unocal	Offshore
13.0	1,924	Texas Pacific	Offshore
37.7	–[a]	Thai Shell	Onshore
–	280	Esso Khorat	Onshore
99.0	3,721		

Note: (a) A small amount of associated gas found at Shell's Sirikit field.

Source: World Bank.

area are high, and World Bank estimates of gas potential in the Khorat region outstrip the estimated potential for the Gulf of Thailand.

There are two important features of gas fields in Thailand which significantly affect the attitude of oil companies towards exploration and development in the country. The first is that the offshore gas is very rich in liquids, providing additional commodities (LPG and condensate), which are easily traded and potentially exportable. The second important feature is the geology of Thailand, which is unfortunately more discouraging to company activity. Fields tend to be complex and broken by many faults, making exploration and development costly, since a larger number of wells are required to find and develop a given amount of gas.

The geological complexity also makes estimation of reserves a difficult process. Total potential reserves in 1985 were put at as much as 20 tcf;[4] but this can only be a very rough guess. Already Unocal has suffered badly through an unduly optimistic estimate of reserves in its offshore field. Reserves of 1.58 tcf in its Erawan field were downgraded to just 628 bcf (40 per cent of the original figure) when production failed to reach the expected levels. The implications of this type of problem are serious for both oil companies and government. From the oil companies' point of view it makes investment in fields far more risky. The reverberations of Unocal's miscalculations in 1978 continued through to the summer of 1986. From the country's point of view, uncertainty over size of reserves makes it extremely difficult to plan a coherent long-term energy policy. For example, power stations that were ready to receive Unocal gas had to suddenly revert back to

[4] World Bank.

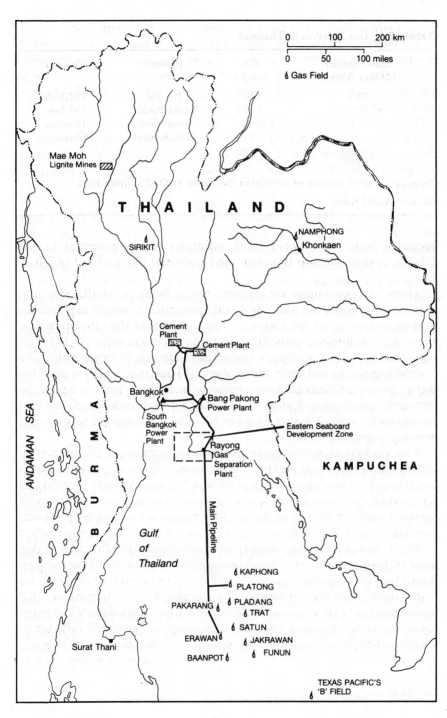

fuel oil when deliveries were much lower than expected. Such uncertainty contributes to an overall reluctance by customers to convert equipment to gas burning, or to make new investments in gas-burning equipment. This in turn feeds back to the oil companies as a very uncertain market, which discourages further exploration and development of reserves.

Gas production from the Unocal field commenced in 1981, with 128 mcf/d. Quantities increased to 214 mcf/d by 1984 and 327 mcf/d in 1985, but fell back to 314 mcf/d in 1986, due to competition from low-priced fuel oil.[5] Details of production problems are given in Section 12.5

(b) *Infrastructure.* The Petroleum Authority of Thailand (PTT) is sole owner of the submarine gas pipeline in the Gulf of Thailand (the longest offshore line in the world at the time it was built). The private sector was not prepared to get involved in this at all, but its construction was an absolute necessity if gas were to be developed. Consequently, the government (through PTT) took 100 per cent of the risk on this $430 million investment. However, some onshore pipelines have been privately financed: namely the 60-mile pipeline from Bangkok to Saraburi, financed by the Siam Cement Company, and the line taking Shell's associated gas from Sirikit to the local power station.

The offshore line is 260 miles long, and has a free-flow capacity of 500 mcf/d expandable to 900 mcf/d with the addition of compressors. PTT also supplied a further 106 miles of pipeline on shore, to the Electricity Generating Authority of Thailand's power plants at Bang Pakong and South Bangkok.

The case of the offshore pipeline is interesting, because PTT sees its willingness to put equity into it as having been a crucial factor in reassuring the oil companies of its serious intent towards gas. In their interpretation, it was not merely an economic necessity that they take on this burden, but just as importantly a moral requirement, which supplied confidence in a market for the gas, not just its transport. PTT succeeded in raising most of the loan financing from the World Bank, together with the Japanese and American Export–Import Banks and commercial banks.

Raising funds from outside sources presented a new problem to the government and oil companies however, and provides an interesting example of the opposing needs of companies and financing bodies, which can leave the government in quite a dilemma. By 1978 both

---

[5] *Ibid*; Hongladaromp, pp. 11, 22.

Union Oil and Texas Pacific had overfulfilled the conditions laid down for the first eight years of their concessions, as regards expenditure and work completed, and they were anxious to go ahead and produce as soon as possible. They felt they 'had done their bit' by proving sufficient gas to meet the initial estimated demand of 150 mcf/d over the twenty-year life of the fields. Their plan was to slow down further exploration, while digesting the massive expenditure required to build production platforms. Drilling operations could subsequently be stepped up again in the final two years of the extension period. While this made economic sense for the companies, the government needed to persuade bodies such as the World Bank to provide finance for a pipeline. And these bodies understandably wanted evidence of proven reserves, not just probable fields. The Thai government therefore insisted that the companies pledge to conduct large-scale additional drilling in the first year of their statutory four-year extension period (1978–82), in order to firm up their reserves. The oil companies retorted that they could only do this in return for a higher gas price, to compensate for the delay in recouping their expenses. The government remained adamant that pricing and renewing the leases were two quite separate issues.

From this point the history of the two companies diverged. Union Oil succeeded in reaching a compromise on prices, and Texas Pacific continued fruitless negotiations over the years. The story illustrates neatly the kind of impasse that can result from the different, but equally legitimate, interests of the two sides in negotiations.

There has been discussion in Thailand over the possibility of privatized pipelines. At the time the major offshore pipeline was built there was virtually no possibility of raising private capital, largely because the pipeline's profitability rested overwhelmingly on the loyalty of a single governmental customer, EGAT. Furthermore, given just one supplier of gas and one major consumer there is actually very little scope for competition; privatization would simply substitute a private for a state monopoly. For the smaller onshore lines private capital has been invested, although in the past ownership has reverted to PTT. For example, in the very early days of gas development the Saraburi cement company decided to build a spur line to its plant, and raised the necessary capital for this investment. Similarly, transport of gas from Sirikit to the local power station is arranged directly between supplier and consumer. A new ruling in 1987 has further liberalized gas infrastructure. Private pipeline companies will now be allowed to lay, own and operate gas lines to serve private sector industrial customers. Ceramics and tobacco-curing, among others, are envisaged as possible customers for such a service.

12.4 The Government Sector

In the past the Thai energy bureaucracy has been criticized for its complexity and lack of direction. The governmental process is characterized by consensus decision-taking, which is unfamiliar to Western oil company executives, who are more used to a clear hierarchical structure. In the energy sector there was a proliferation of committees, scattered in different ministries, each with responsibility for separate but overlapping fields of interest. To some extent they were unified under the Prime Minister's office, and through the same individuals belonging to different groups. But there was a lack of overall co-ordination. The drawing up of the five-year plan was the only occasion when all the various policies were drawn together, and properly discussed in relation to each other.

Such an interlocking system of committees and agencies can help to protect against autocratic decision-making and to check corruption. But it can also paralyse the decision-making process. In response to this problem the newly elected government formed a National Energy Policy Committee in the autumn of 1986. It is too early to judge how well this will perform. Its function is described below, as well as that of its predecessors. In addition the roles of the Ministry of Industry, the National Economic and Social Development Board (NESDB), and the Petroleum Authority of Thailand (PTT) are clarified.

(*a*) *National Energy Policy Committee.* Formed in November 1986 this is designed to take over the functions previously carried out by the National Petroleum Policy Committee and the National Electricity Policy Committee amongst others. It is intended to provide continuity and co-ordination between the various energy sectors in the country, but without going to the extent of forming a specialist Ministry. It is responsible for formulating energy policy; co-ordinating this policy with the ministries and the NESDB; fixing realistic energy prices; and analysing and evaluating the progress of energy-related projects.

(*b*) *National Petroleum Policy Committee.* This was a very large committee, chaired by the Prime Minister, and with representatives from a very wide variety of interests (producers, users, financial, planning, legal, etc.). It suffered from being too large to take useful decisions quickly enough, coupled with its divorce from non-petroleum energy issues. This has now been superseded by the National Energy Policy Committee.

(*c*) *Committee for Negotiation on Petroleum Joint Ventures and Pricing.* This small committee was set up in 1983 with responsibility for overseeing all existing agreements with oil companies and for negotiating new ones.

(d) Ministry of Industry: Department of Mineral Resources (DMR). DMR is responsible for regulating oil and gas licensing, and devising concession terms, royalties, tax rates, etc.

(e) National Economic and Social Development Board (NESDB). This is a national planning agency attached to the Prime Minister's office. Its status is indicated by the fact that its Secretary General is one of just three civil servants who participate in Cabinet meetings. While it is not part of the decision-taking group as such it makes an important input to policies. For example it prepares the five-year plans. Its recent energy plan for 1987–91 appears to be setting the framework for current energy policy in Thailand.

(f) Petroleum Authority of Thailand (PTT). The Petroleum Authority of Thailand was formed in 1978, taking over the role of the former Natural Gas Organization of Thailand (NGOT). It is responsible for gas purchases and sales, gas transmission, and petroleum products marketing. It is thus a classical national oil/gas company, with both upstream and downstream interests. A recent new departure is for PTT to become directly involved upstream, by taking a minority 25 per cent share in a joint venture with Shell. The idea is that PTT should become involved once commercial discoveries have been made; that is, it is not taking on the exploration risk (which is unusually high in Thailand, due to geological conditions). This is seen as a good method to transfer technology to the Thais, giving PTT the opportunity to gain experience in production. The World Bank has provided a loan to PTT for this project, with co-financing coming from commercial banks. PTT is also a joint-venture partner in various downstream operations: the TORC refinery (49 per cent), the National Petrochemical Corporation (48 per cent), the National Fertilizer Corporation (21 per cent), Thai LNG Company (10 per cent), Bangkok Aviation Fuels (10 per cent) and the Bangchak Petroleum Corporation (30 per cent).

PTT for the time being has a monopoly on the purchase and sale of gas, with the exception of Shell's associated gas, sold direct to EGAT. In the past it has also had a monopoly on the distribution of gas, but recent moves have been made to encourage private investment in gas pipelines. Over this issue some rivalry between the NESDB and PTT has become evident. PTT is seen by NESDB as being over-large and over-powerful, and the latter has been arguing that some of PTT's functions should be privatized. Its arguments are couched in terms of 'free-market' ideology, and owe a lot to the voices of the private sector oil companies and institutions such as the World Bank. In the gas arena, NESDB claims that PTT's monopoly on the buying, transmission and selling of gas has led to restricted private investment in gas exploration, development and marketing. This argument seems rather unfair, since PTT's investment in the onshore pipeline was a precondition for Union Oil developing its gas fields. Similarly, with only two major customers for gas (EGAT and the Siam Cement Company) there was little scope for competing suppliers. Furthermore, in the case of Shell's associated gas at Sirikit, PTT is not in fact involved as an intermediary. Nevertheless, the government has accepted privatization as an important strand of the Energy Plan, and the new policy on privately owned pipelines is the first evidence of this with respect to gas.

Over the years PTT has been able to raise significant sums from commercial banks, export credit agencies, multilateral and bilateral institutions, with the help of government guarantees. For example it raised $505 million to finance the offshore gas pipeline. The government is now insisting that it raises money without the backup of a government guarantee for projects such as the refinery expansion, and again this is proving possible. But nevertheless its status as a national enterprise has caused it to suffer from liquidity problems, thus reducing the cash available for investment in new projects. On the one hand it is subject to government directives on its investment policy. For example it is not impossible that the government will require it to invest in a fertilizer plant for political reasons, where it might itself choose not to on economic grounds. On the other hand its major customers are other state enterprises: EGAT (the electricity supply company), Bangkok Mass Transit Authority and the State Railways. Each of these has at times had problems in paying PTT for fuel received. (In May 1986 for example, EGAT owed 1,652 million baht to PTT, BMTA owed 866 million baht and SRT owed 1,278 million baht.[6]) But PTT is effectively obliged to supply fuel to state enterprises whether or not payment is received.

12.5 The Private Sector

Thailand has been open to activities by foreign companies for several decades. During the early 1980s there was a dramatic increase in the

---

[6] *Bangkok Post*, 8 May 1986.

number of exploratory wells drilled; but since the 1986 fall in the oil price, new activity has been very sparse. While this is obviously due in part to outside factors, especially the fall in the oil price, the companies are taking advantage of the opportunity to persuade the government to improve its fiscal and contractual terms for exploration and development. The government recognizes that the international competition for scarce exploration funds is now fierce, and is in fact improving its terms.

A number of companies are involved with gas in Thailand. *Unocal* (formerly Union Oil) has been active since 1968. It found gas at Erawan in 1976, and signed a twenty-year supply contract in 1978. Production started in 1981, and a further twenty-five-year supply contract for gas from Baanpot and other fields was signed in 1982. *Texas Pacific* came to Thailand in 1975, and discovered gas off shore in 1976. It agreed technical details on gas supply by 1979, but pricing discussions were never satisfactorily completed. It is now negotiating the sale of its field to PTT. *Shell* has been active since 1979. It discovered Thailand's only producing oilfield on shore at Sirikit; this has been producing oil and associated gas since 1983. *Exxon* has been in Thailand since 1979; it discovered gas at Namphong in 1981 but price negotiations are still under way. BP and Britoil also have concessions. Marathon, Amoco and Phillips are amongst companies that have withdrawn from Thailand. Petro-Canada International Assistance Corporation has also assisted PTT in exploration over a number of years.

While Unocal succeeded in developing its gas field, and Shell is producing oil with associated gas, no other company has yet moved into production in Thailand. Union Oil and Texas Pacific both discovered their offshore gas fields in the same year, 1976. Both companies were keen to develop the fields, and started price negotiations with the Thai authorities immediately. But whereas Union Oil reached an agreement by 1978, Texas Pacific has not yet resolved its problem. It is therefore of interest to trace the history of each company's activities in the country, and especially their relationships with the government, to determine why this should be the case.

(*a*) *Unocal.* There may well have been corporate differences of approach between Unocal and Texas Pacific. But the crucial point seems to have been that Unocal succeeded in reaching a deal that suited it; once this had occurred the PTT was determined to apply the same well-head price to Texas Pacific, which that company felt was too low to cover its costs. To be the first company to reach a deal was a distinct advantage to Unocal, in addition to the physical features of its reserve which were more favourable than those of Texas Pacific's.

Unocal's price deal, which took two years to formulate, gave it an initial average well-head base price of $1.04/mbtu. An adjustment clause was linked to three variables: the US export price index, the Thai wholesale price index, and the cost of medium fuel oil at Singapore. Unocal's second supply contract, signed in 1982 to cover the additional fields at Kaphong, Platong, Pladang and Baanpot, followed broadly the same formula, although a slightly higher base price was allowed.

Unocal's field has the advantage of being closer to the mainland than that of Texas Pacific. But perhaps the crucial factor which allowed Unocal to reach agreement with the government and to start production in a relatively short time, was the nature of the gas at Erawan. It is very wet gas, with a high proportion of condensates (in fact this proportion proved to be even higher than originally thought). Since LPG and condensates can be exported for hard currency, this gave Unocal considerable flexibility in discussing prices for its methane. In 1985 condensates accounted for 30 per cent of Unocal's total revenues in Thailand.[7] Furthermore, Erawan gas has a high thermal content, which is important since the price is set according to btu rather than volume.

Shortly after production started at the end of 1981, however, Unocal encountered problems. By August 1982 production was only reaching half the agreed flow of 250 mcf/d. Unocal blamed a reservoir shortfall as the main cause, plus technical problems due to complicated and faulty reservoir structures. Unocal claimed that a shortfall in reserves would count as *force majeure*, releasing it from its obligation to deliver the contracted quantities of gas. But the Thais refused to accept this argument; only after a complete new assessment of reserves had been carried out by de Golyer and MacNaughton did they reluctantly accept about twelve months later that reserves at Erawan were only 40 per cent of the figure conjectured by the same firm two years earlier (i.e. 628 bcf instead of 1.58 tcf). In the mean time, the Minister for Industry threatened to cancel the contract if production did not increase by the end of the year. (Both power stations and cement plants had already converted to gas, and were now unable to get supplies.)

It seems Unocal tried to co-operate, by pushing development of the Baanpot field ahead faster than originally planned. But PTT meanwhile brought into play a penalty clause in the gas purchase agreement, which permitted them to pay only 80 per cent of the agreed

---

[7] *Ibid*, 15 May 1986.

price due to shortfalls in delivery. Once PTT accepted the new reserves figure, the price reduction had to cease, with some retro-activity (about $43.5 million). But Unocal had only been paying royalties on the 80 per cent price they had received, whereas the Department of Mineral Resources thought they should have been calculated on the full price. This dispute over royalties finally went to international arbitration in March 1986. There it was ruled that Unocal was in the right, and need only pay royalties on the price it had received from the Thais.

Throughout these problems, the relationship between Unocal and the government remained amicable, and Unocal continued to push ahead with development of new fields. Unocal was able to remain sanguine because other factors were working in its favour. Costs of production equipment were falling, thanks to world-wide cut-backs in oil exploration and production; Erawan gas had a high thermal content (and the price was on a btu basis, not volume); and, most important, there was a larger than expected production of condensate which could be exported. Whereas the original Erawan agreement was negotiated and finalized on the basis of an 18 per cent rate of return, this only came down to 14–15 per cent; without the compensating factors it could have fallen much lower.

More recently the problems of dependency between a single producer and single consumer have been reversed. Unocal is now eager to supply much larger quantities of gas than are specified in the take-or-pay agreement, and more than PTT is willing to buy. The first reason for this situation arising has been the delay in building the planned petrochemical and fertilizer plants. The large gas consumption of these plants was assumed in the company's forecast production levels; already by March 1985 PTT was only consuming 350 mcf/d where Unocal was ready to supply 400 mcf/d.[8] A number of wells were shut in as a result of this shortfall in demand. A far more severe problem arose in 1986 however, with the dramatic fall in the price of oil. With fuel oil available at very low prices the two gas consumers, EGAT and the cement plant, inevitably switched back to burning their original fuel. As a result total consumption of gas fell from 327 mcf/d in 1985 to just 314 mcf/d in 1986. PTT does however forecast a slight rise to 350 mcf/d in 1987.[9] While the government has not defaulted on its contract, the company is disappointed at the lack of opportunity to sell larger quantities of gas, and it has had to shut down 100 out of 250 production wells. Its financial position was

[8] *Oil & Gas Journal*, 18 March 1985.
[9] *Ibid*, 2 February 1987.

worsened by the fall in value of its condensate production by 60 per cent in the first seven months of 1986.

PTT responded to these difficulties in two ways. First it banned all imports of fuel oil, thus restricting supplies to those from domestic refineries. Secondly it reduced gas prices in order to make gas more competitive with fuel oil. Both PTT's share of the consumer price and Unocal's well-head price were cut. In addition the Chairman of PTT (who is also Chairman of EGAT) seems to have wrought from EGAT a commitment to continue using gas to some extent, at least. The government is also trying to encourage continued exploration effort by improving contract terms and fiscal arrangements.

These efforts do seem to have reassured Unocal. Whereas it was claiming in 1986 not even to be breaking even on its $2.6 billion investment, by mid-1987 it is looking for new sales contracts, in order that its fields at Pakarang, Jakrawan, Trat, Funan and North Pladang, will be ready to take over when production from Erawan declines in the 1990s.

(*b*) *Texas Pacific (TP).* The history of Texas Pacific's dealings with the government has been very different. It agreed technical details with the authorities in 1979 on output and pipeline route. But pricing discussions, which started in March 1979, have never been concluded. Negotiations are now taking place over a price at which TP will sell the whole field to PTT and withdraw altogether. TP is anxious to recoup its costs, which are said to amount to $120 million plus interest. Meanwhile several other companies are reported to be interested in the field, should the sale go through.

TP was at a disadvantage relative to Unocal, because the latter was the first to reach a price agreement with PTT, and Unocal was in a position to accept worse terms than TP. The main reasons for this were mentioned above. First, the thermal content of TP's gas was lower than that of the Unocal Erawan field due to high carbon dioxide content, meaning that a greater relative volume would need to be extracted to earn the btu-linked price. Second, pipeline costs would be higher, since the distance from the field to the market was greater. Thirdly, and probably of prime importance, the TP gas contained a much lower proportion of condensates. This meant they could not compensate for a relatively low Thai methane price by selling condensates on the world market. Meanwhile PTT suggested that TP ought to accept a lower price than Unocal, since its gas contained more carbon dioxide, and since TP's reserves were much larger.

An added factor in the negotiations must have been TP's corporate identity and behaviour. TP is a subsidiary of Seagrams, the US/

Canadian distillers. While Seagrams used to own several small oil and gas interests around the world, they have gradually sold all of them over the past few years, with the exception of TP's small foreign interests. This could have had an impact on negotiations with the Thais: a company which is selling all its petroleum assets and is no longer really active in the sector, is not in a good position to convince the last customer that it will be able to serve him.

A further inhibitor to negotiations between TP and the government was the lack of an immediate market outlet for the gas. Union Oil started delivery of gas in 1981; and while there was an initial problem in satisfying demand for the gas, there was general confidence that this was a short-term problem. To diversify suppliers would of course be good policy, but PTT was not able to commit itself to purchases of gas without having specific outlets for it. Furthermore, there would be a capital cost of $150–200 million involved in extending the pipeline to the TP field.

In the face of this problem TP began in 1982 to investigate the possibility of an LNG export project, as a joint venture with the Thai company PSA. This proposal met with a very mixed reaction from the government. The Industry Ministry, backed by the Finance Minister and PTT, strongly supported the idea. In their view, Thailand had surplus gas, and badly needed foreign exchange. However the NESDB, and several cabinet ministers, took a more conservative line, arguing that any 'surplus' gas should be kept in the ground for the benefit of future generations. As a result of these internal conflicts the government temporized, to the frustration of TP which saw its potential export markets disappearing to competitors from Indonesia and Malaysia. Relations were to deteriorate still further later in 1982. The government at last formally agreed to an LNG export scheme, and formed a pilot company, in which the government itself was to hold 60 per cent of equity. The remaining 40 per cent was not awarded to TP. Rather, negotiations got under way with the Japanese, who had expressed interest and who were ready to put up a loan of $56 million. TP management was incensed that, at best, it had been pushed into a minority role, and was in fact on the verge of losing all part in the scheme.

Eventually however the LNG scheme faded out of the picture. One potential market in Korea had been lost to Indonesia. The Japanese had also secured sufficient supplies, and were unlikely to need additional sources of LNG before 1990. It became clear that the Japanese interest had in fact been peripheral, and certainly would have required exceptional terms to make it worth their while. There still remains a possibility of an LNG scheme sometime in the future, but there are no concrete plans.

(c) *Shell*. Shell's operations are concentrated around the Sirikit area in central Thailand, where it owns the country's only producing oilfield. This is a relatively small field (about 30 million barrels of recoverable reserves) and has been in production since 1983. Its current output is about 20,000 b/d of crude together with 25 mcf/d of associated gas.[10] So far this has been sold direct to a local power station (note that PTT has not acted as middle-person in this arrangement). The gas is rich in NGLs however, so a profitable market opportunity is being missed, given the shortage of LPG in northern Thailand. There are plans to build a separation plant at Kamphaeng Phet, to produce 30,000 t/yr of LPG. This $12 million venture will involve Shell as 75 per cent equity holder and PTT 25 per cent.

(d) *Exxon*. Exxon's experience has been similar to that of TP, in that it found sizeable gas reserves but despite lengthy negotiations has been unable to come to a price agreement with the government. Reserves are put at 1.5 tcf, and the gas is 95 per cent methane.

Negotiations have been under way since 1983, and most of the clauses of a production contract have now been agreed, with the major outstanding issue being price. Having already invested $200 million, Exxon has deferred any further exploration until negotiations over pricing are concluded. At the beginning of 1987 a breakthrough occurred however, with both sides agreeing to a new pricing approach. Under this, the gas price would be based on the international fuel oil price, but with a discount to encourage substitution. This discount would increase as the fuel oil price went up. This approach is close to Exxon's preferred pricing method, but the parties are yet to reach final agreement on details of the formula. PTT and Exxon are now thinking in terms of the Namphong field reaching full production of 250 mcf/d in 1991.

12.6 Gas Policy, Contracts and Pricing

(a) *Exploration Contracts and Fiscal Terms*. Exploration activity in Thailand reached a peak in 1983 and 1984, when 109 and 108 wells respectively were drilled. This number fell to just forty-three in 1986 however, and is unlikely to be higher in 1987.[11] According to the local Shell chairman, the contract terms current in 1986 made exploration in Thailand profitable only at an oil price of $25 per barrel or more.

The fall-off in exploration was of obvious concern to the govern-

---

[10] *World Gas Report*, 1 December 1986.
[11] Hongladaromp, 1986, p. 10.

ment, which recognized the need to offer competitive terms to companies. For the world environment not only saw oil companies with very reduced exploration budgets, but also a scene in which many other developing countries had already changed contract terms in order to attract scarce business to their borders. Moreover there was a case for arguing that Thai concession terms had always been rather strict, given the marginal nature of most fields that had been discovered. Consequently, a draft of new terms was proposed in early 1987.

Existing concession agreements were drawn up under the terms of three sets of legislation, which were all written before it became clear that Thailand tends to be a 'high-cost-per-unit-of-production' country. These are:

(a) Petroleum Act, 1971, which lays down concession rules, including provision of a 12.5 per cent royalty tax.
(b) Petroleum Income Tax Act, 1971. This stipulates corporate income tax of 50–60 per cent.
(c) In 1982 the Industry Minister made an Announcement to curb the possibility of windfall profits. This set a 19–20 per cent limit on tax deduction through setting expenses against revenue, and also introduced annual bonus fees on crude oil production. For production of 10–20,000 b/d this fee was set at 27.5 per cent in addition to the normal 12.5 per cent royalty, rising to 43.5 per cent for production of 30,000+ b/d. It should be noted that these clauses were introduced at a time when oil was selling for $34 per barrel.[12]

The net result of these terms was that 40–50 per cent of gross production revenues from a modest discovery accrued to the government. For fields producing more than 30,000 b/d the rate was even higher. Furthermore, because of the clause limiting cost recovery, less profitable fields were taxed at a higher rate than more profitable fields. This was one of the major bones of contention of the oil companies. Companies were anxious that the government take should be based on profits, and not on straight revenues or volumes of production. Pressure in this direction also came from the World Bank.

The new draft proposals changed most of these provisions:

(a) It would replace the fixed 12.5 per cent royalty with a sliding scale, from 5 per cent on 2,000 b/d to 15 per cent above 15,000 b/d.
(b) It would eliminate the minimum state take of 27.5 per cent in new reservoirs.

[12] *The Nation*, 24 November 1986.

(c) There would be a 35 per cent windfall profits tax on earnings over $200 per square metre over the whole of an exploration tract.
(d) The fee to preserve rights in tracts not being actively explored would rise fiftyfold.
(e) Onshore and offshore producers would be treated equally.
(f) The right to take disputes to international arbitration would be limited.

While some of these suggestions were welcomed by the companies, others are clearly less attractive to them. It will be important to ensure that in the final package new drawbacks do not outweigh the incentives on offer to companies.[13]

(b) *Pricing*. While PTT and Unocal succeeded in 1981 in negotiating a gas price which suited them both, the changing scene in oil prices has undermined its applicability. Furthermore, the overall energy pricing policy in Thailand has come under severe criticism from the NESDB internally and the World Bank externally. Consequently, interim solutions have been found to the immediate problems, and in its current Energy Plan the government is planning a major overhaul of its basic principles of energy pricing.

The base price for gas, set at the beginning of Unocal's contract, was established on a cost-plus basis. Under new arrangements it would be more likely to be set on a market basis, in relation to the prices of competing fuels. (The recent price cuts do in fact approximate this.) According to the traditional formula this base price was then varied according to an adjustment formula (the trade calls it an 'escalation formula', which in the light of recent events is clearly a misnomer). The adjustment formula included three parameters: fuel oil prices (weighted at 40 per cent), a Thai price index (about 30 per cent) and a US price index intended to represent changes in the price of world exports (about 30 per cent). The values of the parameters were applied with an average time-lag of eighteen months. A foreign exchange conversion factor was then applied, to translate the price from dollars to baht.

PTT sought an eighteen-month time-lag mainly because it was negotiating at a time when oil prices were expected to continue to rise. As buyer it naturally thought that it could gain by delaying the escalation. It was also worried about its ability to immediately pass on price increases to EGAT, since government policy was to keep electricity prices as low as possible. PTT was thus trying to protect itself

---

[13] *World Gas Report*, 10 April 1987.

from being squeezed between rising gas prices and fixed electricity prices. However, although the long time-lag is advantageous to the buyer when prices are rising, it obviously has unfavourable effects when prices fall. In 1986 PTT found itself in a quandary. It was paying a gas price to Unocal based on the relatively high fuel oil prices of eighteen months earlier. At the same time it had to try and persuade EGAT to buy this gas, when the latter could quite easily revert to burning fuel oil, which was readily available at much lower prices. While EGAT was free to buy its fuel where it chose, PTT was left with three unpalatable options. First it could simply accept that it would make a loss, for an unlimited period. Secondly, it could renege on its lifting contract with Unocal, with incalculable consequences for its future relations with the oil industry. Thirdly, it could try and persuade Unocal to accept a new pricing formula.

Eventually PTT succeeded in the latter, in what represents a radical departure from the strict contract terms between Unocal and PTT. PTT has reduced its margins on transmission of the gas, and also agreed to reduce the well-head price. Nevertheless, this solution is seen as a temporary measure, leaving the basic contract untouched. PTT professes satisfaction with the basic terms of the contract. It is even committed to reimbursing Unocal's lost revenue, should it be able to recoup it in future through charging higher prices to EGAT. Another interesting feature of events in 1986 has been the emergence of differential pricing to various customers. For example, while the price to EGAT was reduced to 70 baht, Siam Cement was offered this price only if it consumed at least 15 mcf/d. Other industries were to be assessed on a case-by-case basis.

Although PTT has managed to salvage the situation, it is clear that the eighteen-month time-lag is unsatisfactory, despite its attraction in times of rising prices. A three-month lag would be more reasonable. The case of Thailand also illustrates the problems that beset those trying to devise a pricing system when the market is administered at one end and subject to market forces at the other. Inevitably distortions arise.

12.7 Markets

In October 1986 314 mcf/d of gas was consumed. Of this amount, 74.2 per cent went direct to EGAT (the power authority), 22.3 per cent to the gas separation plant, and 3.5 per cent to the cement, glass and ceramics factories.[14] Including the gas which is treated at the

[14] Hongladaromp, 1986, p. 22.

separation plant, nearly 98 per cent of the methane produced in 1983 was used to generate electricity.[15] The cement industry was originally the second main customer for Unocal's gas; but as a private enterprise it has been even more inclined than the state-controlled power authority to switch back to low-priced fuel oil since 1986. The third major market for gas may in future be provided through the various heavy industries planned as part of the ambitious Eastern Seaboard Project. First envisaged in 1980, this project has been subject to considerable political manoeuvring and is still the subject of lively political debate. Other market prospects which are being actively pursued include an increase in consumption by small industries, the possibility of substituting CNG for diesel, and the further extraction of LPG from the rich gas for use in the domestic, transport and light industry sectors.

(a) *Power.* Initial plans for gas utilization focused almost entirely on the power sector, with the aim of replacing expensive imported fuel oil. This was also seen as the easiest way of creating a market for gas. Three 300 MW steam units at the South Bangkok power station were converted from fuel oil in time for the first gas supplies from the Gulf in 1981. A series of large gas turbine units (8×60 MW) were also commissioned at Bang Pakong by mid-1981. A number of additional units have since been added at Bang Pakong. By 1985 these two power stations were capable of utilizing 400 mcf/d of gas. This dramatic change in generating capacity is shown in Table 12.2. Total installed capacity in June 1985 amounted to 6,305 MW, of which 3,478 MW was steam, 1,809 MW hydro, 720 MW combined-cycle, 265 MW gas turbine, and 34 MW diesel.[16] Only minor additions to this capability are planned over the next twenty years, however, with combined-cycle units being given preference over gas-fired steam plants.

Despite government plans to concentrate gas use in the power sector, several problems have been encountered. The first was caused by the early failure of Unocal to produce gas at the expected rate. Although the supply problems were overcome after two or three years, a vital part of the market had been lost in the mean time. Some stations merely reverted to using fuel oil, for which they had been originally designed. But EGAT also responded to the shortfall in gas by investing in lignite-fired power stations. This is not without its advantages, since it diversifies the nation's power sources while still using indigenous natural resources. The episode did however damage

[15] World Bank.
[16] NESDB, p. 104.

Table 12.2: Electricity Generation in Thailand.

% Generated From	1981	1984
Fuel Oil	64.89	29.9
Hydro	17.04	17.7
Lignite	10.47	10.2
Natural Gas	1.18	38.3
Diesel	1.43	0.3
Imports from Laos-Malaysia	4.99	3.6

Source: NESDB, *Energy Issues and Policy Directions in the 6th National Economic and Social Development Plan (1987–1991)*, pp. 102–3.

the reputation of gas, which was a new fuel to Thai industry. The early investments in converting power stations proved premature and disappointing, even though they were later put to good use.

An associated problem with devoting gas resources to the power sector proved especially acute in Thailand. While the plan to substitute for imported fuel oil made economic sense, in order to save scarce foreign currency, it neglected the fact that 53 per cent of fuel oil consumed in 1980 came from Thai refineries.[17] Thus, while the rapid penetration of gas into the power sector was a success from one point of view, it was also creating a surplus of fuel oil, which could not be profitably exported due to depressed world markets.

This problem has become even more acute since the 1986 slump in oil prices. PTT has completely banned the import of fuel oil, in an attempt to maintain some market share for gas. But nevertheless fuel oil from Thai refineries has recaptured a large part of the power market, simply due to its low international price. PTT has countered this by reducing gas prices to a level equivalent to fuel oil. But it remains to be seen how long term an effect this will have.

According to the current Energy Plan there is potential for increasing gas burning to 925 mcf/d by 1995, from 350 mcf/d in 1985,[18] if the TP field is taken into account. This plan includes the use of gas in two additional power stations, at Nan Phong and Bang Prakong. Quantities of gas available from currently producing Unocal fields are likely to decline by 1995, but Unocal is hoping to bring additional fields on stream.

(*b*) *Cement and Other Existing Industries.* In 1986 gas consumption by

[17] World Bank.
[18] NESDB, p. 105.

industry was just 11 mcf/d, a considerable decrease on previous years. For example the cement industry alone used 25 mcf/d in 1984, and only 5 mcf/d in 1986.[19] Once again, this is a symptom of the very low fuel oil prices of 1986.

The Siam Cement Company was in fact one of the very first major users of gas in Thailand, financing its own spur line to its plant at Saraburi, 95 miles north-east of Bangkok. Originally they planned to use as much as 45 mcf/d, to replace fuel oil.[20] The actual history of their usage runs parallel to that of the power sector. There were early problems when Unocal failed to deliver sufficient quantities of gas, and more recently they have reverted to burning fuel oil on simple grounds of price. Since the cement company is privately owned PTT has far less influence over its fuel choice than in the case of EGAT. However PTT has been trying to give Siam Cement a price incentive to use gas. In the recent rounds of price cuts (October 1986) the price of natural gas was cut from 83 baht to 70 baht per mbtu (equivalent to the price of fuel oil) on condition that Siam Cement used at least 15 mcf/d. At lower consumption levels the price would be reduced by a lesser amount, to 77.70 baht per mbtu.[21]

Other industries using gas as a fuel include the ceramics and glass industries. Each consumer is relatively small in this sector however. In the latest price cuts PTT has negotiated each deal individually, on the assumption that the price cannot be as low as for Siam Cement, which is a bulk purchaser.

PTT is trying to encourage smaller industries to switch from burning LPG to gas, on the grounds of both cost and convenience. It is said that about twenty factories are looking into this possibility. Collectively they could consume about 30 mcf/d.[22]

(c) Planned Industries: the Eastern Seaboard Development Plan. This ambitious plan first surfaced in 1980. Official policy at that time saw the use of gas as a fuel to be wasteful: it was likened to burning teak wood to produce charcoal. EGAT was therefore seen as a major consumer only until such time as other higher-value uses came on stream by about 1987. Industries would then receive gas in priority over EGAT, even if this meant the power sector reverting to fuel oil as gas reserves became depleted.

The plan was to develop a series of heavy industrial plants using

[19] Hongladaromp, 1986, p. 22; World Bank; *The Nation*, 4 October 1986.
[20] World Bank.
[21] *The Nation*, 4 October 1986.
[22] Hongladaromp, 1986, p. 22.

gas as feedstock, along the route of the existing gas pipeline from Rayong to Bangkok. The new development was intended to fulfil two aims: first, to make the best economic use of available gas, and second to promote economic development of a region outside the metropolitan area of Bangkok. There were already deep-sea port facilities at Sattahip, but these would be extended and two additional deep-sea ports constructed. Other infrastructure such as railways and water supplies would also be needed, as well as additional housing.

Total investment needed for the scheme was estimated at $3,800 million.[23] The gas-using components would be as follows:

(a) A gas separation plant, to retrieve the valuable LPG component from Erawan gas and provide the feedstock for two major new industries.

(b) A petrochemical complex which would use ethane and propane from the separation plant to produce 300,000 t/yr of ethylene and 73,000 t/yr of propylene. These would constitute the feedstock for five downstream units.[24]

(c) A fertilizer complex, requiring about 27 mcf/d of gas in addition to imported sulphur and phosphates.[25]

From the very early days the Eastern Seaboard plan has been controversial, since some factions felt that gas should in fact continue to be used mainly in the power sector. The economic viability of the scheme has been endlessly debated, along with all other possible motivations of the scheme's proponents. For example it has not gone unnoticed that the Japanese have been particularly keen to promote it, at a time when they have had surplus capital looking for an outlet. Mingled with the basic economic argument has been vigorous discussion over the level of government involvement in the scheme. The consensus now seems to be that the government's role should be limited to provision of the necessary infrastructure, with private capital bearing the risk of what should be commercially viable enterprises. The debate is still vigorous over both the fertilizer and petrochemical complexes. The only part of the plan to have come to fruition has been the gas separation plant, with which everyone agreed. In January 1987 however a contract was awarded to a Japanese consortium to build the ethylene-propylene complex (a $180 million turnkey project). This is now scheduled to be completed by December 1989.[26]

[23] *Ibid*, p. 14.
[24] Eastern Seabord Development Committee.
[25] National Fertilizer Corporation.
[26] *Platt's Oilgram News*, 8 January 1987.

Gas Processing Plant. This plant opened in early 1985, with a capacity to process 350 mcf/d of natural gas. The methane content then goes to substitute for fuel oil in power and industry. These other products are also produced:[27]

(a) LPG, 450,000 t/yr. This goes to meet internal demand in the domestic, transport and light industry sectors.
(b) Natural gasoline, 66,000 t/yr. This goes to the oil refineries for gasoline.
(c) Potentially, ethane, 350,000 t/yr. This would be used as a raw material in the petrochemical industries, but is not being produced until the plants are built.

Total costs of the LPG plant were $320 million.[28] 64 per cent of this sum was raised from Japanese sources, with most of the remainder coming from the World Bank and the Commonwealth Development Corporation and PTT's internal cash.

The government has had an active policy of promoting the use of LPG – as a substitute for imported diesel and gasoline in the transport sector, and as a substitute for steadily depleting fuelwood and charcoal in the domestic sector. The extraction of LPG was therefore a high priority, and the Rayong separation plant has proved a great success. PTT was keen to expand its capacity to process a further 200,000 mcf/d. Government approval has now been given to add this new unit, at a cost of $77 million. It will produce a further 234,900 t/yr of LPG and 47,000 t/yr of natural gasoline.[29] It should come on stream in 1988.

A gas separation plant will also be built in early 1988 at the Sirikit oilfield. This will proceses 25,000 mcf/d and will produce about 30,000 t/yr of LPG, from gas which is currently being used unprocessed in the local power station.[30] The location of the field is very convenient, since it is in one of the remoter parts of the country, where the government particularly wishes to encourage LPG use, but where distribution is usually quite difficult. This plant is to be built with Shell's 75 per cent participation.

Petrochemicals. The National Petrochemical Corporation is 49 per cent owned by PTT, with the IFC holding 9 per cent equity, and four downstream companies sharing the remainder. It was due to construct a $280 million upstream complex in 1986, but this was post-

[27] Hongladaromp, 1986, pp. 15–16.
[28] Petroleum Authority of Thailand, p. 3.
[29] *Oil & Gas Journal*, 23 March 1987.
[30] *Platt's Oilgram News*, 8 April 1987.

poned due to poor market prospects. Recently the contract has been awarded. The intention is that it should receive feedstock from the gas separation plant, in the form of ethane and propane; in turn it will supply ethylene and propylene to a downstream complex. The latter, comprising four groups of factories, would require investment of $370 million.[31] This entire sum would need to be raised from the private sector.

The project has been delayed due to poor market prospects. The export market is depressed, and the local market is also severely limited, since demand within Thailand is still low by international standards. Annual consumption of plastic products is about 4 kg per person, compared with 20 kg in Taiwan and South Korea, or 30 kg in the USA and Japan.[32]

Fertilizers. Controversy still rages around this project. One month it is given the go-ahead and the next called off, only to be reactivated shortly afterwards. The ultimate decision will depend on the final balance between political and economic arguments. The most recent official government attitude has been to wash its hands of it, treating it as a completely private concern, which will stand or fall on its own merits. Nevertheless this does not seem to be a unanimous view within the government. For example, the Industry Minister made the following statement in October 1986: 'Supposing that I'm an investor, I will not be interested because of the project's low return on investment, but as a politician, I have to support it as it would benefit the farmers.'[33]

Demand for fertilizers in Thailand is relatively low at present, but is expected to grow. Currently all fertilizer is imported, so if the plant is built it is intended to supply the internal market almost exclusively – the Thais will not be looking for export markets for the fertilizer. Some groups are even arguing that the products should be subsidized, in order to encourage faster growth in use of fertilizer and therefore in agricultural productivity (hence the Industry Minister's reference to helping the farmers). The economic argument is over the rationale for building such a plant, when imported fertilizer is available at low prices, and loans will in any case need to be serviced in hard currency. The critics' case is strengthened by the examples of nearby plants in Indonesia and Malaysia, which have been running at a loss, despite gas prices of only $0.60 per mbtu, compared to the Thai price of $2.[34]

[31] Hongladaromp, 1986, p. 19.
[32] World Bank.
[33] *Bangkok Post*, 15 October 1986.
[34] *The Nation*, 17 October 1986.

The total cost of the project is put at $365 million, for a plant with a total capacity of 1 million tonnes per annum (16 per cent urea and 63 per cent NP, NPK).[35] The equity holders include the government at 33.3 per cent, the construction company at 10 per cent, the IFC at 12 per cent, and Thai private investors at 21.7 per cent. Loan finance is also on offer from the Japanese Overseas Economic Cooperation Fund (OECF) and from the IFC. But a major set-back to the project has been the appreciation of the yen, since the proposed contractor is to be paid in yen with Japanese (OECF) financing. The fact that investors are still eager to be involved in the project was cited by the ex-Chairman of NFC as proof of its sound base. Others however have pointed out that several of the investors are in fact suppliers of either equipment or phosphate rock.

There is now a strong vested interest in seeing the project come to fruition, despite the vociferous opposition from some quarters. Nevertheless, the Chairman of NFC resigned after a heart attack in the autumn of 1986, and within a month his replacement was publicly regretting that he had accepted the position. He called it 'the most stupid decision of his life'.[36] The same journal quoted an unidentified official as condemning the project in the following terms: 'the raw material is imported, the energy is effectively imported because we pay for the gas in dollars, the construction is imported, and the technology is imported. The only local content is the stupidity.' The outcome of this debate is still to emerge.

(*d*) *Residential and Commercial.* The use of gas as a primary fuel in the residential and commercial sectors is not high on the agenda. Traditional fuels such as charcoal and wood have largely been replaced by LPG and electricity – which account for more than 70 per cent of cooking in urban households. Electricity is the most important commercial energy used in urban households (about 45 per cent of the total). Kerosine (33 per cent) is used for cooking in restaurants and hotels.[37]

(*e*) *Transport.* PTT is considering the use of CNG as a transport fuel. A pilot project, using CNG in buses, was carried out in Bangkok with the assistance of New Zealand experts in this field. As in other developing countries the market for transport fuels is highly concentrated. The Bangkok area accounts for 80 per cent of total cars and taxis, 70 per cent of buses and almost 30 per cent of trucks. This

---

[35] Petroleum Authority of Thailand.
[36] *The Nation*, 24 November 1986.
[37] World Bank.

would be an encouraging environment into which to introduce CNG.

However, the market prospects do not seem very hopeful. Historically, government policy has been to use price differentials to encourage consumers to move away from gasoline and towards LPG and diesel. The policies have been dramatically successful, and created considerable problems. Thailand now has a surplus of gasoline, produced by its own refineries, which cannot be profitably exported at today's prices. Meanwhile virtually all taxis in Bangkok run on LPG. In 1984 94 per cent of new light and passenger trucks produced were diesel fuelled. Between 1980 and 1984 consumption of LPG increased by 28 per cent a year; similarly consumption of diesel increased by one-third between 1982 and 1984, accounting for 68 per cent of all fuel in the transport sector in 1984.[38] As a result 60 per cent of all imported petroleum products are diesel, and Thailand also needs to import LPG, despite increased indigenous production from the gas separation plant. Especially while oil prices remain low there is therefore an argument for vehicles to convert back to use of gasoline, which is in surplus, rather than to CNG. At most, CNG could only be expected to penetrate small carefully chosen sectors of the market, and would not absorb large amounts of natural gas.

12.8 Finance

Raising the capital for gas projects is not a major problem to Thailand. It enjoys the confidence of the international financial community, and also benefits from the special interest of Japan, which considers South East Asia to be of strategic importance. The country's reputation, and its comparatively low debt/service ratio (amongst the lowest of all the middle-income countries) are a product of the government's extremely prudent approach. A priority in government policy has been to keep the national debt as low as possible. There is a Committee for National Debt, chaired by the Minister of Finance, which scrutinizes all potential loans, in both the public and private sectors.

For example, in a recent case, funds were on offer for a refinery project from the Japanese OECF without a government guarantee. But the government still hesitated for eighteen months over its decision to accept the loan. In the case of the proposed fertilizer project it is said in some quarters that the main problem is not the economic viability of the plant, but the decision to take on more foreign debt.

Unlike the poorer developing countries there is also private capital

[38] *Ibid.*

available within the country, which is a further advantage. For example, private investors are committed to 21.7 per cent of the equity in the fertilizer plant.

The future of gas in Thailand falls squarely on the economic benefits which it can offer to the country. If satisfactory projects can be identified then finance is available. If such projects still do not materialize then the reasons must be with either government or oil company policies, or the interaction between the two. In the case of Thailand these relationships have not always been easy, and there is scope for criticism of pricing policy and economic planning. But ultimately the major problems with gas utilization have arisen from external sources, and most especially the collapse in the oil price. This left both government and companies in a very difficult situation. It is to the credit of both sides that relationships are still relatively good; that gas is still flowing; compromises have been accepted; and the future for gas in Thailand still looks hopeful.

Sources

Eastern Seabord Development Committee, *Thailand's Petrochemical Complex, Prospectus*, Bangkok, December 1982.

Hongladaromp, Tongchat, *Natural Gas Development in Thailand*, PTT, October 1986.

National Economic and Social Development Board, *Energy Issues and Policy Directions in the 6th National Economic and Social Development Plan (1987–1991)*, Bangkok, October 1985.

National Fertilizer Corporation, *National Fertilizer Complex*.

World Bank, internal documents.

Much of the information which lies behind this chapter was obtained in interviews with executives from the following companies and organizations: BP, Exxon, Foster Wheeler, Government of Thailand: Ministry of Industry and NESDB, Petroleum Authority of Thailand, Shell, Texas Pacific, Unocal and the World Bank.

APPENDIX TO CHAPTER 12

Background Data

Population[a]: 51.7 million
GNP per capita[a]: $800
Average annual growth in GNP per capita, 1965–85[a]: 4 per cent
GDP[a]: $38,240 million

External public debt[a]: $13,268 million
External public debt as percentage of GNP[a]: 36

Total energy consumption[b]: 17.238 mtoe
  Commercial: 77 per cent   Non-commercial: 23 per cent
Commercial energy consumption per capita[a]: 343 kgoe
Commercial energy imports[b]: 10.671 mtoe
Commercial energy exports: n/a

Proven Reserves and Production[c]

	Gas	Oil	Coal	Hydro
Proven Reserves	3,721 bcf	99 mbls	865 mt	10,050 MW
Production	400 mcf/d	35,000 b/d	3.7 mt	3,660 GWh

Consumption by Sector: Thousand Tons of Oil Equivalent[b]

	Gas	Oil Products	Coal	Total	% Share
Power	1,346	1,802	368	3,534	27
Industry	33	1,705	180	1,918	15
Commerce	–	236	–	236	2
Transport	–	5,657	–	5,657	44
Households	–	447	–	447	3
Agriculture	–	433	–	433	3
Other	–	681	–	681	5
Total	1,379	10,961	566	12,906	
% Share	11	85	4		100

Power Sector[c]

Installed capacity: 6,305 MW
Electricity generation (%): Hydro=18, Oil=30, Coal=10, Gas=38

Sources

(a) World Bank, *World Development Report*, 1987, pp. 202–237. Figures refer to 1985.
(b) World Bank, internal reports. Figures refer to 1983.
(c) National Economic and Social Development Board, *Energy Issues and Policy Directions in the 6th National Economic and Social Development Plan (1987–1991)*, Bangkok, October 1985, pp. 4, 5, 86, 101–4. Figures refer to 1985, except for coal production and electricity generation shares which refer to 1984.

13 TUNISIA

13.1 Country Summary

Tunisia belongs to the ranks of 'middle-income' developing countries, thanks largely to its exports of oil and phosphates and to tourism. These resources allowed the country to develop at quite a fast rate through the 1970s; but price falls for both export commodities have caused severe problems to the economy in recent years. In the case of oil this problem has been magnified by the fact that known reserves are steadily depleting. Tunisia expects to become a net importer of oil, probably as soon as 1990.

In response to this situation the government is trying to encourage greater exploration and development of its hydrocarbon resources. On the exploration front it introduced new terms for joint-venture agreements in 1985, in an effort to attract more companies to its shores. The new incentives included two features of great importance to oil companies. The ring fence was removed, and the gas price was linked to the international price for fuel oil. Unfortunately however these reforms were made just before the collapse in oil prices, after which they no longer made such an impression on companies. The good intentions of the government were undermined by outside events beyond its control. With regard to development of oil and gas fields the government is pursuing the possibility of the national oil company, ETAP, developing further its own expertise, to enable it to take on the development of small or marginal fields which are of little interest to foreign companies, but could be of crucial significance to the nation. It has also explored the possibilities of developing the large offshore gas reserve at Miskar. Unfortunately however the latter would be a very costly exercise, and for the time being it seems to have been pre-empted by the 1986 collapse in oil prices.

The situation of Tunisia is in some ways unique however. Traditionally its medium-term fortunes were seen to depend on finding an alternative fuel to the fast depleting oil reserves. Discussion focused on the choice between indigenous natural gas and imported coal, both of which were quite costly solutions. Both possibilities have nevertheless been studied in detail, both by the government and by organizations such as the World Bank. The experiences of 1986 however have shown that Tunisia can benefit considerably from its geographical position, to obtain imported energy at less than the normal international prices. This had not previously been taken fully into account. Tunisia has exceptionally easy access to both internationally traded gas and fuel oil, without the need for long-term commitments to either energy source. First, Algerian gas passes through the country in the trans-Mediterranean pipeline; the Tunisian government earns a royalty on this service, which is payable in either cash or kind, and which can be supplemented by further purchases if necessary. Second, tanker loads of 'distress' fuel oil in the Mediterranean spot market can also be purchased by Tunisia on favourable terms whenever the need arises. No transport costs are incurred on Algerian gas (other than internal distribution), thus reducing costs; and since fuel oil requirements are small the government does not need to enter into long-term contracts. It can choose the lowest-priced cargos as and when they are required. These lower than normal import prices are an important bonus for Tunisia, and make it doubly important that over-costly indigenous resources are not unwisely developed.

Although the immediate urgency for gas development has been alleviated through making use of such strategies, the government is nevertheless very interested in exploiting the gas reserves within its territory, providing this can be achieved in an economically acceptable way. The most hopeful prospects for the medium term lie with its small reserves, which could use the existing infrastructure, rather than with the large Miskar field which would require capital investment out of proportion to the country's size and resources.

13.2 The Energy Balance in Tunisia

Commercial energy consumption in Tunisia has been heavily dependent on oil. In 1984 97 per cent of commercial energy came from hydrocarbons, with 83 per cent from oil and just 14 per cent from gas.[1] The balance was provided by hydroelectricity and a very small amount of imported coal.

Oil has been one of the major products of the country; in the early 1980s oil exports provided as much as 50 per cent of total merchandise exports (although prior to 1982 crude oil was also imported, to take advantage of the higher quality of Tunisian crude, so that net exports of oil accounted for just 22 per cent of exports).[2] It fuelled an annual

[1] Ministère de l'Économie Nationale, 1984, p. 4.
[2] Dinh, 1984, p. x.

growth rate of 6–7 per cent per annum through the 1970s, and while internal consumption increased considerably with economic development, production also rose gradually to meet this demand. The producing fields are now approaching depletion however, and few new fields have been discovered. Proven remaining oil reserves (in fields already developed) are only 320 million barrels,[3] enough to provide another ten years of production at current rates. Exploration efforts have tended to be unfruitful, and in a climate where companies were in any case cutting their exploration budgets, this left a bleak outlook for oil in Tunisia. But recently Marathon discovered a good oilfield at Tzarzis, and this may encourage further exploration. Failing further discoveries however, the country is expected to be in energy deficit by 1990.

Ever since 1971 natural gas has been used to a limited extent as a fuel in Tunisia. Consumption grew very fast following the opening of the Algerian trans-Mediterranean pipeline, with the share of gas increasing from 15 per cent of the total energy mix in 1980 to as much as 30 per cent in 1985. However it fell back to just 20 per cent in 1986, reflecting the availability of cheap fuel oil in that year, and the government's ability to switch easily from Algerian gas to imported oil.[4]

Looking to the future the government can choose between importing coal to meet its energy needs, or developing its gas reserves. A decision between these alternatives was probably imminent at the time of the oil price fall, but has now been postponed. The difficulties in developing the indigenous gas will be made apparent in later sections.

13.3 Gas Reserves, Production and Infrastructure

(a) *Reserves and Production.* Proven remaining gas reserves in Tunisia, at the beginning of 1986, were 1.8 tcf.[5] The two fields producing gas are both on shore: Cap Bon in the north of the country close to Tunis, and El Borma in the south-west, on the Algerian border. (Some of the El Borma gas is in fact produced from Algerian territory, and marketed through Tunisia.) Cap Bon has been supplying gas since the 1950s. The oldest producing oilfield, at El Borma, has been providing Tunisia with associated gas since the early 1970s (99 per cent of the

---

[3] ETAP, direct communication.
[4] *Ibid.*
[5] Information in this section was obtained from ETAP and from unpublished World Bank sources.

gas consumed came from El Borma and the nearby field Chouech Es Saida, and 1 per cent from Cap Bon). Both fields are now ageing however, and both oil and gas production are decreasing: oil production at El Borma is expected to fall from 72,000 b/d in 1984 to 30,000 b/d in 1994. Remaining reserves of gas at El Borma are put at about 1 tcf.

The largest single gas field is the offshore reserve of Miskar, which was discovered by Elf-Aquitaine in 1975. Reserves here were originally thought to be as high as 2 tcf, but this figure was later reduced to 1 tcf proven, with a further 1 tcf of probable reserves. The gas is non-associated, but there are geological difficulties in retrieving it. Furthermore the gas is sour, with nitrogen and carbon dioxide making up one-third of its volume.

A number of smaller fields have also been found, both on shore and off shore. Onshore finds have been at Djebel Grouz and at El Franig. Djebel Grouz, near El Borma in the south-west, was discovered by Pecten in 1981, and is now under the aegis of a partnership between AGIP (60 per cent) and Pecten (40 per cent). Recoverable reserves are estimated at 70 bcf, with potential production of 10.5–14 mcf/d. It has not yet been developed, pending confirmation of the reserves and decisions over use of spare pipeline capacity (see Section 13.5). El Franig was discovered by Amoco, in an undeveloped interior region close to the Algerian border. Originally Amoco was very optimistic about its potential, but reserves have since been reduced to an estimated maximum of 140 bcf of gas, with 20 million barrels of condensate. Furthermore the field is 100 miles from the main market, the industrial area on the coast around Gabès. Nevertheless Amoco is still considering its development. Off shore Marathon has two small gas fields, at El Biban and at Bregat. Reserves are estimated at 30–40 million barrels of oil, with 60 bcf of gas, plus some gas with condensates.

When looking at Tunisia's gas reserves mention should also be made of the Algerian pipeline which passes through the country, before crossing the Mediterranean to Europe. This line was opened in 1983, and under the initial contract Tunisia was entitled to 5 per cent of the gas throughput in royalty payments for the twenty years of Algeria's contract with Italy. This amounted to 21 bcf a year, or the cash equivalent. The Société Tunisienne de l'Électricité et du Gaz (STEG) also negotiated a separate deal to purchase additional gas from the Algerians: 2 bcf between April and September 1985, rising to 5 bcf in 1986. The purchase agreement ended in 1986, and the Tunisians were hoping to replace this with an interruptible supply agreement thereafter.

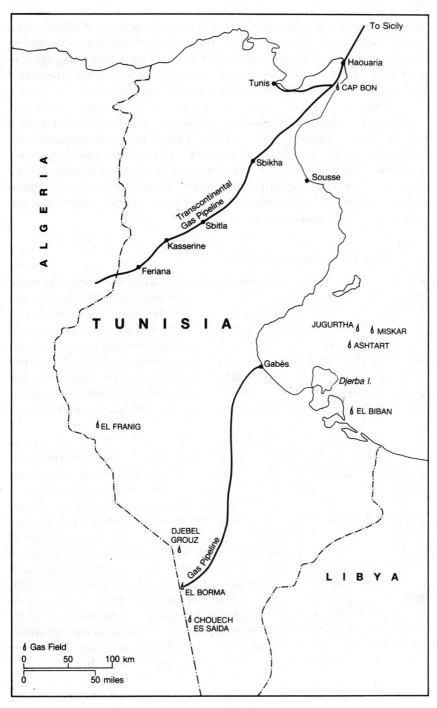

To Sicily

Haouaria

Tunis

CAP BON

ALGERIA

Sbikha

Sousse

Transcontinental
Gas Pipeline

Sbitla

Kasserine

Feriana

TUNISIA

JUGURTHA ◊ ◊ MISKAR

◊ ASHTART

Gabès

Djerba I.

◊ EL BIBAN

◊ EL FRANIG

DJEBEL
GROUZ
◊

Gas Pipeline

LIBYA

◊ EL BORMA

◊ CHOUECH
ES SAIDA

◊ Gas Field

0   50   100 km

0   50 miles

(*b*) *Infrastructure.* Distribution of gas is the monopoly of STEG. The first pipeline was built in 1971, financed by the World Bank, to transport gas from El Borma to Gabès. This line is still being used to its full capacity, with the possibility that gas from Djebel Grouz will take up any slack, as El Borma becomes depleted.

In 1983 the trans-Mediterranean pipeline from Algeria to Italy, which passes through Tunisia, was opened. This provided a golden opportunity to Tunisia, and with World Bank financial support it was made the basis of a national network. Four more lines were constructed in the north of the country, at a cost of $44.8 million. The total network now amounts to 610 miles. More pipelines are planned for the south of the country, but these will only be built if the gas reserves in that area are developed.

13.4 The Government Sector

(*a*) *Ministries.* Prior to 1986 there was no Ministry for Energy as such. Responsibility for the oil and gas sector was divided between the Ministries of National Economy and of Planning. The Ministry of National Economy set both well-head and retail prices, and prepared the five-year plans; but the Ministry of Planning was responsible for finalizing the sectoral five-year plans. This division of responsibilities made it very difficult to co-ordinate policies, and to ensure that the different interest groups worked towards the same end. This sector has now been reorganized into one Ministry for Energy and Mines.

(*b*) *Entreprise Tunisienne d'Activités Pétrolières (ETAP).* ETAP acts as joint-venture partner with foreign oil companies. It co-ordinates exploration and production activities, and all imports of crude and products. Traditionally it has been very much a sleeping partner in exploration activities, putting up minimal amounts of risk capital, and leaving all responsibility for programmes to the foreign companies. Recently however there have been indications that the government wishes to change this, and to develop the capabilities of ETAP as an independent operator. This is partly in response to the oil companies' complaints that ETAP is avoiding all the risk involved in exploration. But it is also seen as one way in which development of small or marginal fields, of little interest to foreign oil companies, may be facilitated. If the Tunisians were able to make full use of their small oil and gas fields they might be able to avoid, at least temporarily, the huge investments needed to develop the large Miskar field.

(*c*) *Société Tunisienne de l'Électricité et du Gaz (STEG).* STEG is responsible for the production, transmission and distribution of electricity and gas. The entire gas pipeline network therefore falls under its remit. There is an obvious logical reason for combining the administration of gas and electricity when 98 per cent of gas is used to generate electric power; it can also be argued that in organizational terms the distribution of gas is a very similar operation to the distribution of electricity. However there is some uneasiness over the resultant concentration of political power in STEG. It has been recommended in the past that the two functions, gas and electricity, be devolved into separate institutions, in order to avoid this concentration. As yet this has not happened however.

13.5 The Private Sector

All oil and gas developments are carried out through joint ventures between foreign oil companies and ETAP, with the exception of the El Borma field which was developed by SITEP (50 per cent government, not ETAP, and 50 per cent AGIP). Prior to the 1970s the Tunisian oil scene was largely the domain of European, and especially French, companies. While there is still a strong European presence, approximately half of the companies now present in Tunisia are American, attracted by the stable political and economic environment.

The peak in exploration activity came in 1982, reflecting renewed enthusiasm after the oil crises of the 1970s, and encouraged by the discovery of the Ashtart oilfield in 1971. Since then results have not been outstanding however, and interest has fallen back to a lower level. In the last ten years fifteen small oil deposits and ten small gas fields have been discovered, but only five oilfields have gone into production, and many dry wells have been drilled. The geology of the country is generally acknowledged to be complex, making exploration a highly risky activity. Nevertheless nearly all the available acreage is under exploration agreements, and a slow-but-steady rate of activity continues.

Only two of the gas fields have been developed, El Borma (with Chouech Es Saida) and Cap Bon. The government has for some years been keen to use more gas, but policy has alternated between encouraging the use of Tunisia's numerous small fields, and trying to launch the large Miskar project. The position has been further complicated by negotiations with the Algerians over purchase of gas from the trans-Mediterranean pipeline. For example, it has been suggested that the enthusiasm for developing Tunisian gas fields has sometimes been simply a negotiating ploy to reach better terms with the Algerians. Even if this is not the case, the fact that some oil companies perceive it in this light must add difficulties to their discussions with the government. The current position seems to be that the Miskar project is indefinitely stalled, due to a combination of high capital costs and the low fuel oil price. Negotiations over the smaller fields continue, but there seems little immediate likelihood of progress, given the low price of alternative fuels.

Djebel Grouz is the subject of active negotiations between Pecten, AGIP and the government. The field lies 30 miles from El Borma, and only 9 miles from the main gas pipeline which runs from El Borma to the industrial centre of Gabès. In the past the capacity of the line was saturated by associated gas from El Borma, but by the end of 1986 some spare capacity was predicted, as production at El Borma fell off. This is obviously an ideal situation in the sense that little additional infrastructure is required, but there is still disagreement between the companies and the government over implementation of the scheme. ETAP wants the companies to simply take up the spare pipeline capacity, and use the existing separation plant at Gabès. The companies on the other hand want to build a small treatment plant near the field, thus enabling them to fit more gas into the pipeline.

Marathon is also negotiating with the government, over development of its fields at El Biban and Bregat. The company is generally satisfied with the returns it is able to earn in Tunisia, but in the case of these gas fields little progress has been made. Marathon wants to recover past exploration costs through charging a relatively high price for gas, and through an inexpensive development scheme. Given the geological complications however this will mean that only 5 per cent of the oil and 35 per cent of the gas are recovered. Neither the price nor the recovery factors are acceptable to the government.

Amoco has relinquished all its exploration permits, but still holds the El Franig production concession. There are no plans for its development in the near future however.

Tunisia's largest gas field at Miskar was discovered by Elf-Aquitaine in 1975, and ever since then has been the subject of lively debate. In 1977 a decision was taken to go ahead with its development, with projected gas flow rates of about 200 mcf/d by 1981. The Groupe d'Étude Miskar (a joint Franco-Tunisian body) was set up to co-ordinate the project, and bids were taken for hardware, including a 220 mcf/d treatment plant on shore near Sfax. The scheme then lapsed, following the sinking of a disappointing well in 1979. Two years later, in 1981, the government tried to revive it. SEGMA (Société d'Études de Gaz Marin) was formed, with participation from both Elf and the government (40 per cent each) and Apicorp (20 per cent), with the task of studying the potential of all the offshore fields.

They recommended shelving the Miskar project, mainly because of the lack of either an export or internal market at that time. Market problems were compounded by the poor quality of the gas, and the very high capital costs involved. The Tunisian government saw an alternative fuel in cheap imported coal. Meanwhile Elf had better investment opportunities in other countries, and objected to the proportion of risk capital that was expected of it in Tunisia. Elf subsequently relinquished the field. Opinion has remained divided ever since over the economics of the project. The consensus seems to be that at an oil price of $15 the project should be viable; and a study looking at the possibility of limited recourse financing for it judged that an acceptable rate of return could be earned on a capital investment of $500 million. The major problem however lies in raising that capital sum. And given the availability of cheap fuel oil the government seems to have lost the impetus to struggle to raise it. The scheme is therefore lying in abeyance for the time being.

13.6 Gas Policy, Contracts and Pricing

Companies have not been very active in exploring in Tunisia in recent years, and newcomers have been discouraged by the fact that a number of small fields have remained undeveloped. As a result the government became anxious to encourage greater activity. Following consultation with the companies themselves and with the World Bank they brought out new regulations in the autumn of 1985, aimed especially at facilitating the development of smaller and/or less profitable fields. The major clauses of these regulations are as follows:

(a) There is accelerated depreciation up to a maximum of 30 per cent per annum.
(b) Whereas in the past exploration costs were imputed for each separate concession, costs from one concession can now be recouped when developing other concessions in the same permit. This is of special help for marginal or expensive areas.
(c) Taxation and royalties are calculated and paid on a sliding scale, according to revenues and costs.
(d) The government has made a commitment that the gas price should be equivalent to 85 per cent of the international price for heavy fuel oil.[6]

(*b*) *Pricing.* Before the new exploration terms of 1985 there was no

---

[6] *Petroleum Economist*, May 1986.

formal gas pricing policy. In practice domestic prices were aligned with the heavily subsidized fuel oil prices, which brought them to below the opportunity cost. This caused an obvious problem for companies potentially interested in developing gas reserves. Over recent years however the government has progressively increased energy prices, and following the fall in international oil prices, domestic prices are now more or less at parity with international levels. As the government has now also committed itself to setting gas prices at 85 per cent of the international level for fuel oil the companies have a much firmer basis for negotiation.

13.7 Markets

Consumption of gas in 1984 amounted to 75.5 mcf/d, with just over half of this quantity (40 mcf/d) coming from the Tunisian fields, and the remainder from the Algerian pipeline.[7] The share of gas in total commercial energy consumption rose from 15 per cent in 1980, to 30 per cent in 1985, falling to 20 per cent in 1986 thanks to the fall in price of fuel oil.[8]

The bulk user for gas has always been the power sector. There was discussion of a fertilizer plant in 1982, but this came to nothing, which was probably realistic given the market situation. The premium residential and tourist sectors have been targeted for connection to the distribution network, although their total consumption is unlikely to be very large.

(*a*) *Power.* 98 per cent of gas consumed in Tunisia goes to the power sector (79 per cent to STEG, and 19 per cent to industrial energy plants). The proportion of electricity that is generated from gas has grown very fast, to reach 80 per cent in 1985. In that year just 15 per cent was generated from oil, with the remaining 5 per cent from hydro. The potential for hydroelectricity in Tunisia is relatively small, at 350 MW.[9]

Table 13.1 shows how fast the generating capacity in Tunisia has grown over the last twenty years, and how gas has taken up a large proportion of that growth.

(*b*) *Industry.* Industry has provided the main market for natural gas, outside the power sector. Few small industries use gas however: 80 per

---

[7] *Petroleum Economist*, January 1986.
[8] ETAP, direct communication.
[9] *Ibid.*

Table 13.1: Electricity Generation in Tunisia. 1962–85. Megawatts.

	1962	1972	1982	1985
Steam	70	228	527	790
Gas Turbines	–	15	319	489
Hydro	28	28	28	64
Diesel	19	19	4	4
Total	117	290	878	1,347

Source: *Revue Tunisienne de l'Énergie*, July 1986, p. 65.

cent of all industrial consumption is accounted for by just ten large users. These include the cement industry, mining and chemicals. Debate still continues whether gas is the most economic fuel for these industries. For example a study has recently been completed by a Canadian company (financed under a Canadian aid programme) which looks at the possibility of converting all six cement plants to imported coal.

(*c*) *Residential and Commercial.* Tunis itself had a town gas supply long before natural gas became available. This meant that a distribution network was conveniently in place once the natural gas became available. The existence of a flourishing tourist sector, in the same neighbourhood as a wealthy residential district made the supply of natural gas a worthwhile proposition. Concentrations of tourist hotels in Sousse and Cap Bon have led to these towns also being connected to the national grid.

13.8 Finance

In the past the government succeeded in promoting projects in the oil sector where virtually all the risk capital was provided by the oil companies. This era has now come to a close however. The companies are no longer willing to continue under these conditions, and the government itself is anxious that ETAP develops its own capability to carry through projects by itself. ETAP itself does not have the capital available to finance such moves however, and it is possible that the World Bank would become involved if ETAP were to go ahead with development of a small field for local use.

The other main focus of financial concern in the energy sector has been the means of raising capital for the proposed development of Miskar gas. Elf has not shown very much enthusiasm for the scheme, which leaves the government with the task of raising approximately $400 million from aid and/or the commercial money market. The major difficulty is the size of this scheme in relation to the overall budget of the country, since it represents almost 60 per cent of Tunisia's total investment budget. This inevitably adds greatly to the risk of a project which already has an uncertain market.

The World Bank has paid much attention to this problem, and has explored at length the possibility of organizing commercial limited recourse financing for it, with guarantees provided by itself. This could effectively circumvent the problem of the size of the investment in relation to the country's resources. These studies, carried out before the collapse in the oil price, judged that Miskar would earn an acceptable rate of return on the capital invested. Special long-term low-interest loans would not be required, providing the gas were sold at approximately 85 per cent of the international heavy fuel oil price. It found that banks were hesitant about such an investment mainly due to fear of the political risks of currency convertibility and offtake commitments, not because of the project's lack of intrinsic merit. The World Bank's suggestion was that it could guarantee these aspects of the investment: if the government defaulted then the World Bank would step in to pay the company. The government would then owe the money to the World Bank rather than to the commercial banks, with three months' grace for payment. Given its dependence on the World Bank in other sectors of its economy it could not afford to miss repayments.

This idea seemed like a hopeful route forward; but unfortunately it has been superseded by the oil price collapse, which has led the whole Miskar project to be shelved. The scheme was not without its problems however, notably the probable reluctance of the government to risk loans in the agriculture and other sectors on a gas development scheme. Furthermore, as outlined the scheme does not allow for the fact that the government might default on uptake commitments for quite legitimate economic reasons. It is not necessarily reasonable for the oil company to expect to pass all the market risk on to the government. Lastly, it could have put the World Bank in the invidious position of having to arbitrate between the company and government, and even having to enforce an uneconomic policy on the government.

For the foreseeable future the level of risk associated with developing Miskar is too high in relation to the benefits it could bring to Tunisia. The energy needs of Tunisia are quite modest and can be adequately and economically supplied from other sources. Ideally these sources would include some of the country's smaller gas reserves; but its easy access to Algerian gas and to fuel oil tankers plying the Mediterranean should not be underestimated. This strategy will also leave Tunisia with a resource which may prove to be of much greater value to future generations than it is today.

Sources

Dinh, H.T., *Oil and Gas Policies in Tunisia: A Macroeconomic Analysis*, World Bank Staff Working Paper 674, 1984.

Ministère de l'Économie Nationale, Direction Générale de l'Énergie, *Problèmes, Options et Orientations Énergetiques à l'Horizon 2000*, 1984.

World Bank, internal papers.

Much additional information was obtained from discussions with executives in the following companies and organizations: AGIP, Amoco, Elf-Aquitaine, ETAP, Marathon, Ruhrgas, Shell, and the World Bank.

APPENDIX TO CHAPTER 13

Background Data

Population[a]: 7.1 million
GNP per capita[a]: $1,190
Average annual growth in GNP per capita, 1965–85[a]: 4 per cent
GDP[a]: $7,240 million

External public debt[a]: $4,688 million
External public debt as percentage of GNP[a]: 59.2

Total energy consumption[b]: 3.690 mtoe
   Commercial: 93 per cent   Non-commercial: 7 per cent
Commercial energy consumption per capita[a]: 546 kgoe
Commercial energy imports[b]: 2 mtoe
Commercial energy exports[b]: 4.1 mtoe

Proven Reserves and Production[b]

	Gas	Oil	Coal	Hydro
Proven Reserves	1.8 tcf	1.8 bbls	–	65 MW
Production	18 bcf	40 mbls	–	3,570 GWh[c]

Consumption by Sector: Thousand Tons of Oil Equivalent[b]

	Gas	Oil Products	Coal	Total	% Share
Power	804	206	–	1,010	29
Industry	189	768	70	1,027	29
Commerce	9	178	–	187	5
Transport	–	831	–	831	24
Households	4	302	–	306	9
Agriculture	–	163	–	163	5
Total	1,006	2,448	70	3,524	
% Share	28.5	69.5	2		100

Power Sector[c]

Installed capacity: 1,347 MW
Electricity generation (%): Hydro=3, Oil=20, Coal=0, Gas=77

Sources

(a) World Bank, *World Development Report*, 1987, pp. 202–237. Figures refer to 1985.
(b) ETAP, direct communication. Figures refer to 1985.
(c) *Revue Tunisienne de l'Énergie*, No. 5, July 1986.

INDEX

Aba, 173
Abu Dhabi, 4, 98n
Abu Gharadiq, 132, 134, 137
Abu Madi, 129, 132, 134, 137, 143
Abu Qir, 132, 134, 137, 144
Afam, 167, 173, 179
Afghanistan, 4, 98n
AGIP, 9n, 26, 137, 138, 172–3, 175, 181, 207,
    211, 216, 217, 256, 259
Agrico, 217, 218
Aid agencies, 13, 17, 27, 47–8, 70, 90, 92, 94, 99,
    111–12, 125–6, 142, 144, 182
Ajaokuta, 167, 170, 180
Akukwa, 169
Aladja, 167, 170, 180
Alakiri, 167
Alexandria, 132, 134
Alfonsin, R., 111, 113, 114, 118
Algeria, 4, 98n, 138, 254, 255, 256, 258, 262, 264
Alif, 132
Amerada Hess, 35
Ammonia, *See* Fertilizers
Amoco, 26, 35, 50, 116, 117, 137, 139, 195, 211,
    233, 256, 260
Amoseas, 172
Amu, L., 171
Angola, 8
Ankara, 6
Apicorp, 260
Appraisal of discoveries, 85
Argentina, 4, 17, 19, 20, 33, 37, 40, 45, 48, 89,
    98n, 103–28
Ashdod, 144
Ashland, 172, 175, 181
Ashtart, 259
Asian Development Bank, 192, 201
Associated gas, 5–6, 21, 43, 62, 106, 132, 137,
    151, 164–5, 167, 172–3, 196, 256
Athens, 6
Atlantic Richfield (Arco), 35, 137
Attock Oil, 194
Austral basin (Argentina), 105, 106, 109
Austria, 218

Baanpot, 234
Badin, 188, 189, 191, 192, 194
Badr el-Din, 132, 137, 139
Bahía Blanca, 105, 111, 113, 122
Baluchistan, 193
Bangkok, 228, 242, 244, 245, 248–9
Bangkok Mass Transit Authority, 232
Bangladesh, 4, 20, 98n, 217
Bang Pakong, 228, 242
Bang Prakong, 243
Bank Bumiputra, 155
Banks, 48, 74, 89–90, 104, 111, 125
Beersheba, 144

Belayim, 132
Belgium, 218
Bendel, 168
Benin City, 170
Biafra, 174
Bintulu, 154, 160
Biomass, 13, 20, 215, 246
Bolivia, 4, 98n, 104, 105, 108, 116, 119, 120, 123
Bonny, 181
Brazil, 4, 33, 37, 98n, 104, 109, 123–4
Bregat, 256, 260
British Gas, 26n, 143, 180, 195
British National Oil Corporation (BNOC), 70
British Petroleum (BP), 26, 39n, 137, 139, 155,
    174–5, 179, 180, 181, 189, 195, 211, 233
Britoil, 233
Buenos Aires, 104, 108, 109, 111, 113, 119, 121,
    122, 123
Bureaucracy, 99, 123, 126, 134, 192–3, 230
Burma, 4, 98n
Burmah Oil, 26, 31, 50, 186, 191, 193, 194, 195,
    196

Cairo, 20, 26n, 134, 143
Caltex, 32n
Campo Durán, 105, 109, 124
Canada, 59, 263
Cap Bon, 255, 256, 259, 263
Cement, 18, 215, 242, 243–4, 263
Central Luconia, 150, 155
Ceramics, 18, 244
Chase Manhattan, 89
Chevron, 172
Chile, 4, 98n, 105, 123, 124
China, 4, 13, 98n, 99, 191, 201, 218, 219
Chouech Es Saida, 256, 259
Chukai, 158
Coal, 11–12, 19; infrastructure, 12, 142; prices,
    12
Cogasco, 104, 110–12, 125
Colombia, 4, 98n
Commerciality, 85, 86, 88
Commercial sector, 19–20
Committee for Negotiation on Petroleum Joint
    Ventures and Pricing (Thailand), 230–1
Commonwealth Development Corporation, 218,
    246
Commonwealth Development Finance
    Company, 193
Commonwealth Secretariat, 210, 212, 213
Compressed natural gas (CNG), 8, 13, 22–3,
    122, 144, 159, 178, 200, 209, 216, 248–9
Condensates, 23, 132, 137, 151, 226, 234, 236,
    256
Consultants, 87
Conch, 180
Conoco, 155, 195